C000220046

BRITISH TRACTORS

Benjamin Phillips

NOSTALGIA ROAD

In memory of
Tom & Jean Bradley, my Grandparents,
and Ken Briggs, neighbour
and good friend

First published 2017

ISBN 978 0 7110 3858 5

Published by Crécy Publishing Ltd

Printed in Poland by Opolgraf

Nostalgia Road is an imprint of
Crécy Publishing Limited
1a RingwayTrading Estate
Shadowmoss Rd
Manchester
M22 5LH
www.crecy.co.uk

Picture credits
Every effort has been made to identify and correctly attribute photographic credits. Should any error have occurred this is entirely unintentional. All uncredited pictures are from the author's collection.

FRONT COVER **A couple of New Holland T Series tractors working in the British countryside.** *Case New Holland (CNH)*

Left picture **Field tests of the Nuffield tractor were carried out in Great Britain in 1946 at a time when tractors were in great need. This half-tracked version is fitted with a trailed plough.** *Nuffield and Leyland Tractor Club*

Middle picture **Siromer has always kept its tractor designs up to date, adding to their popularity.** *Clinkaberry Tractors*

Right picture **A Turner undergoing tests, probably in Wolverhampton around 1949.** *turnermanufacturing.org.u*

BACK COVER **Top picture A very early John Deere row crop tractor, this un-styled Model A hasn't even got rubber tyres fitted.** *John Deere*

Second picture **HNP 417, a 1949 Ferguson TEA20 belonging to the author's father, Peter Phillips. It was restored to its present condition in 1991.** *Peter Phillips*

Third picture **A David Brown Cropmaster. The gentleman leaning on the bonnet is the author's good friend Malcolm Cooper, who is now in his late 80s.** *Jean and Malcolm Cooper*

Last picture **Many manufacturers today make tractors that run on tracks. This is Case IH's version, a Quadtrac 620.** *Case New Holland (CNH)*

Contents

About the author

I have been interested in tractors since a very early age, when I would watch them work in the field next to our garden in Trimpley. I restored HNP 417, a grey Ferguson T20, with my Dad in the summer of 1991, aged 12. Just over a decade later I started restoring tractors for other people and have traded under the name The Tractor Lad ever since. In 2006 I started writing for *Tractor and Farming Heritage* magazine and have been a regular contributor to that publication.

LEFT **A very young Benjamin Phillips, interested in tractors from an early age.**

Acknowledgements

A big thank you goes to the following people.

My Mum and Dad (Kate and Peter Phillips) for putting up with lots of tractors about the place over the years and their help and support. Also to my Dad for supplying his archive pictures throughout this book and buying the Ferguson tractor that started all this.

Amy Trevitt, my partner, who has helped me with everything, including this book – your support is always greatly appreciated.

Thanks also to Jean and Malcolm Cooper; Jane Brooks, Peter Squires, Ross Bartlett; John Deere; The Nuffield and Leyland Club; Dave Sanders (Machinery Decals.co.uk); Ted Everett (AGCO Archives); Case IH; New Holland; Clinkaberry Tractors; Turner Manufacturing.org.uk; Claas UK; JCB; Kubota; Steyr; Renault UK.

Introduction

One of the things we are lucky to have in Great Britain is lush green countryside which produces great farm land where we can grow a variety of crops. Along with fertile farm land we need good agricultural machines and Great Britain has over the years produced some of the best farm tractors in the world. In this book I have covered most of the popular models that have either been built or just used in this country. Although some of the manufacturers in this book aren't British, such was the size of the market that they had big facilities situated around the country which became famous. Places such as Basildon, Bradford, Coventry and Doncaster all became synonymous with tractor production and provided employment for thousands of people. Other tractors included in the book may have been popular in certain regions and possibly not seen in others; this is largely due to the locations of different dealerships which would have had a significant influence on their distribution as many farmers were extremely loyal to one dealer and would buy only from their trusted dealership.

In the book you'll see tractors from the beginning of their development when they were relatively simple machines (some even without tyres) up to their present-day equivalent. Fordson and John Deere tractors were prominent in the early years until Harry Ferguson, an Irish inventor, developed a system that was to revolutionise tractor design. Deciding on a factory in the centre of the country, in Coventry, he built probably the most famous tractor ever produced – the grey Ferguson. The Ferguson TE20 laid the foundations for how tractors would be designed right up to how we know them today. Of course today tractors rely on electronics to do everything and looking inside a cab you'd be forgiven for being confused by the amount of switches there are. However, no matter how complicated tractors have become over the years, they still have the three-point linkage in common with the little Ferguson which was built in Coventry.

Unfortunately, very few tractors are made in Great Britain today and the famous factories have been replaced with housing estates or shopping centres. In 2002 the Banner Lane factory in Coventry which was home to Massey Ferguson closed its doors for good as production relocated to France. The departure of such a big brand as Massey Ferguson from the UK signalled that the heyday of tractor production in this country was over. This left the New Holland factory in Basildon, Essex, as the only tractor plant from the 1960s still in use and it celebrated 50 years of production in 2014. Despite there being a vastly reduced tractor production in Great Britain today, the market here is still as important as ever and all the major tractor brands import thousands of models every year. As this book covers both tractors made and used in Britain, it has allowed me to include a whole raft of new models that weren't made within these shores but have been seen here in great numbers. Great Britain will always be an important country for tractor manufacturers as we will hopefully always be blessed with rich farm land, but we have also played a big part in tractor development over the years and this should never be forgotten.

Benjamin Phillips
2017

Allis-Chalmers

It is unsurprising to find that Allis-Chalmers was started by an entrepreneur by the name of Edward P. Allis. In 1860 he obtained at an auction a bankrupt factory in Milwaukee, Wisconsin, where he quickly set about improving the fortunes of this ailing business. He soon saw that the need for steam power was ever increasing, so deciding to build steam engines seemed a good choice. However, in 1873 there was financial panic in North America, and even further afield in Europe, which hit the company that Allis had bought quite hard. Luckily his good reputation saved him and he was just about able to weather the crisis. His steam engines were used predominantly to power flour mills, so when he employed well-known people in this field the company prospered greatly. Even more success came when Edward's two sons, Charles and William, took over on their father's death in 1889.

At the turn of the 20th century things were going so well that the company was now producing more steam engines than probably anyone else in America. In 1901 the company merged with that of two Scottish immigrants, Thomas Chalmers and Thomas Dickson, and from then on was known as the Allis-Chalmers Company. For the next 20 or so years Allis-Chalmers built a whole array of machines to deal with a form of extractive metallurgy; in fact anything to do with metal and smelting was its forte. It also produced a whole host of mills.

BELOW **The Allis-Chalmers Model U was in production from 1929 to 1952.** *Peter Phillips*

LEFT Adopting a bright orange colour made other manufacturers follow suit. This Allis Chalmers WF was in production between 1938 and 1951. *Ben Phillips*

BELOW Allis-Chalmers tractors really took off in Great Britain with the Model B, introduced in 1937. *Ben Phillips*

Now in need of more help, Otto Falk arrived to assist in making the company more financially sound. He saw that agricultural machinery was set to expand greatly and wanted the company to capitalise on this. It therefore started to build tractors and implements, mainly for the American market at first. Bright colours made the tractors stand out, so Allis-Chalmers painted theirs in Persian Orange, and this move made other tractor manufacturers follow the 'bright colour' idea.

Allis-Chalmers really took off in Great Britain when the company introduced the Model B in 1937. This small orange tractor was simple, manoeuvrable and light, so the appeal was universal and many farmers quickly took to it. For many years these

LEFT The D270, introduced in 1954, was basically an updated Model B. *Ben Phillips*

tractors were imported from the USA, but in the mid-1940s a production shortage prompted the company to open a factory in Britain. A site in Southampton was chosen, but it was with the factory the company later moved to, in Essendine, Lincolnshire, that Allis-Chalmers tractors became most associated. The Model B was still a popular tractor; it was best for row crop work as the crops were visible from the driver's seat and the whole tractor had a naturally high ground clearance. In addition, the British versions had an adjustable front axle. The 22hp petrol paraffin engine was quite capable of doing a number of jobs; then later some tractors benefited from having a Perkins P3 diesel engine fitted, together with a basic hydraulic system. Amazingly the Allis B was in production in the USA and Great Britain for 18 years and did not alter in shape or size at all.

The next Allis-Chalmers model to emerge from Essendine was the D270, which was similar to the B, with the Perkins P3 and petrol paraffin engines carried over. However, by the mid-1950s the more economical Perkins diesel was more commonly offered, being the preferred engine option and gaining a good reputation. The D270 also had updated tinwork, which looked slightly different from the B. Most of the long steering column, so distinctive on the Model B, was now covered up by an awkward-looking section of tinwork that contained the battery, which, with the simple dash, did little for the appearance of the tractor. The popular Allis B was in production for 18 years, but the D270 only managed three.

In 1957 the Allis-Chalmers D272 entered production with better styling, but still based on the basic design from the B with the same three engine options of petrol, vaporising oil and diesel. However, the constant updating on the original Allis B platform was doing the company no favours, and it was falling behind the market leaders in Great Britain. A whole new model was therefore called for.

It is strange to think that Allis-Chalmers thought it could compete with the likes of Massey Ferguson when in 1960 it launched the ED40, with an engine that Massey had discarded just a year before. If you mention the 23c diesel engine built by the Standard Motor Company to most tractor enthusiasts they will probably screw up their faces. It was first fitted in the Ferguson FE35 and gained a terrible reputation for its poor starting ability; farmers who bought a Ferguson with the 23c were soon tearing their hair out when it wouldn't start. Ferguson ditched this

RIGHT The section of tinwork housing the battery and dash on the D270 looked awkward. *Ben Phillips*

unpopular engine for a Perkins, which was ironically an updated version of the one found in the D272 that Allis-Chalmers had replaced. The 23c used in the Allis ED40 was fitted with a heater plug under each injector to try and improve the poor starting, which helped the situation but it was still far from perfect.

The ED40's styling was largely borrowed and slightly adapted from the American-built D series, and with so much of the new tractor taken from other models it was hardly the new model customers wanted. At the end of the 1960s ED40 production stopped, as did tractor production at the Essendine factory, and Allis-Chalmers disappeared from Great Britain for good. Although it continued to produce tractors in America for many years after this, in the early 1990s, after many different owners, Allis-Chalmers formed AGCO (Allis Gleaner Corporation), which now owns Massey Ferguson, Fendt, Valtra and Challenger, and is one of the biggest tractor manufacturers in the world.

ABOVE **Thousands of the Model B tractors were imported into or made in Great Britain between 1937 and 1954. Many are still about today and can be seen lined up like this at many shows in Great Britain.** *Peter Phillips*

ABOVE Entered into production in 1957, the D272's looks were a big improvement on the D270. *Ben Phillips*

LEFT The ED40 was not a great success and when production stopped in 1968 it signalled the end for Allis-Chalmers in Great Britain as this was the last model produced at Essendine in Lincolnshire. *Ben Phillips*

TOP AND BOTTOM RIGHT Some badge-engineering was in evidence, as these Allis models actually started off as Massey Ferguson tractors that were made in Great Britain. In America the Allis name was more popular. *AGCO Archives*

8765
AGCO ALLIS

8745
AGCO ALLIS
2010 stubble drill
tye
tye

Austin

ABOVE **The Austin Model R tractor was made in Great Britain and France, but couldn't compete with the Fordson in Great Britain, so production moved solely to France where it continued until 1951.** *Peter Phillips*

Many people think that Austin only made cars, but in 1919 an Austin Model R tractor rolled off the Longbridge production line. Was it a coincidence that the design was very similar to the Fordson F7 Maybe, but many believed that the Austin was of better quality than its familiar-looking rival. However, it was more expensive and had unproven reliability, so the Fordson rival was more popular.

Painted in dark blue with red wheels and having the name 'Austin' in the famous signature script (which went on to adorn many a car grille) in bold across the side of the bonnet, this tractor certainly looked very desirable. With its little 20hp petrol/paraffin engine, the Austin tractor was exported far and wide, ending up in Australia, South America and Africa; it was also exported to France, where it became more popular than in Great Britain. The French Government placed high costs on imported goods, preferring the purchasing of French-made products, so to get round the problem of the added financial burden imposed on its tractors Austin opened a factory in France, at Liancourt, just north of Paris. Parts from Great Britain were taken across the Channel and the tractors assembled at the new factory. This was a shrewd move that worked in Austin's favour in two ways: it was able to make a British product that the French farmers really loved at a reduced price, and the Fordson F, which was made in either Cork, Ireland or Dearborn, USA, was still constrained by the high import cost. As the Longbridge factory started to produce the famous Austin 7 car, tractor production moved fully to France, continuing there until the early 1950s.

This little unassuming British tractor helped Austin through difficult times after the First World War and paved the way financially for the company to produce the cars for which the Austin name became famous throughout the world.

BMB

John Brockhouse was born in Wednesbury, Staffordshire (now West Midlands), on New Year's Eve in 1844. Twelve years later, as an orphan, he began an apprenticeship with an axle-maker, and over the next few years stayed in the same industry, moving from company to company within the West Midlands area. In 1886 he was working for John Rigby & Sons, having gone there aged 15, then had a brief spell working for Richard Berry & Sons before moving back to Rigby. When John Rigby died, Brockhouse knew that the business would probably flounder, so set up his own company behind a butcher's shop, making parts for the automotive industry. The Black Country area was renowned for making heavy metal items such as axles and springs, in which Brockhouse specialised. The company flourished, with up to 10 men in full-time employment when it moved to a bigger site in West Bromwich. Brockhouse started to make a wide range of products, and during the First World War times were good. Brockhouse himself carried on working in a managerial capacity right up to the age of 79, then his son, also John, took over.

Producing foundry work for a number of companies was the key to the company's success; supplying Aga cookers as well as the British railway and shipbuilding industries allowed Brockhouse to become a major player in drop-forging. By the mid-1940s there were 24 companies in the Brockhouse group, and in 1947 the addition of the British Motor Boat Manufacturing Company allowed it to bring out the BMB President tractor.

Although made by Brockhouse, the BMB President used the British Motor Boat company initials. Introduced in 1947, the tractor used spare parts left over from the Second World War and was fairly simple. However, it was never going to appeal to customers with big farms as the engine was not very powerful; it was a 10hp four-cylinder Morris industrial unit at just under 1000cc, and ran on petrol TVO (tractor vaporising oil). Horticulturalists and smallholders were the main users of the President; it was good amongst crops as it was light, the steering was offset, and vision along the crop rows was good. The rear wheels were adjustable to make row-cropping easier; this was done by slackening the clamp

BELOW **The little President tractor was made from 1947 to 1956 by Brockhouse in West Bromwich, the West Midlands.** *Dave Sanders*

on the wheel hub and sliding it across a square shaft; the latter was around knee height and could be quite painful if you forgot to avoid it as you walked round the tractor!

A couple of variants of the President came in the guise of vineyard and orchard models. To achieve these more specialist models the upright exhaust was changed so that it exhausted from beneath the front of the tractor. The air cleaner mushroom was removed from the other side of the bonnet; this was not particularly high, but every bit of clutter was tidied up. The orchard version, which would probably need to travel under low fruit trees, had the same exhaust and air cleaner modifications as found on the vineyard variant, and it also sat 13 inches lower.

By 1956 sales of the President were poor, so production ceased; another firm bought up the remaining parts and fitted a couple of different engines, naming the result the Stokold tractor.

Today the BMB President is popular with a number of collectors largely due to its size and low weight, which makes it easier to store and transport. The company that produced this little tractor is still in business and still in Howard Street in West Bromwich, where it has been since 1888.

TOP LEFT This capable little tractor featured a basic lift system. It couldn't lift heavy objects but would work the land with ease weeding rows with mid- and rear-mounted tool frames. This picture was probably taken in the late 1940s. *Dave Sanders*

BOTTOM LEFT Mid-mounted implements could be fitted, providing a useful addition to the tractor. Here, in this picture taken in the late 1940s, a salesman is demonstrating the virtues of this little tractor. *Dave Sanders*

RIGHT Today the BMB President is popular with collectors due to its light weight and manoeuvrability. *Ben Phillips*

BELOW Many collectors use the President today for ploughing matches. *Ben Phillips*

Case

In 1842, when Jerome Increase Case took an everyday threshing machine and changed its design to make it better, he probably never imagined that the company he would later set up bearing his name would become one of the biggest tractor manufacturers in the world.

From an early age Jerome had an interest in agriculture, especially the harvesting side. His family were farmers and he would see them using all manner of machines to harvest wheat and other cereal crops. This interest led Jerome to move to Wisconsin, where there were better opportunities to sell the kind of machines he was interested in. When he arrived he bought a number of threshing machines and immediately saw the problems that beset them; after he had made improvements they were a great sales success. He needed to give this new company a name, so decided on the J. I. Case Company, and so this famous brand was born.

With Jerome at the helm the company was in a strong position and he was very well respected in Wisconsin, being appointed a mayor on two occasions. There were many reasons behind

ABOVE LEFT The Case DEX was introduced to Great Britain as part of the Lend-Lease scheme. *Peter Phillips*

BELOW LEFT Case adopted this bright flambeau red after many years of drab colours. The LA model on the right dates from 1940-1953; the small tractor on the left is a Case S, built from 1941 to 1952. *Peter Phillips*

ABOVE When Case owner Tenneco bought International Harvester, both names appeared on its tractors. *Ben Phillips*

J. I. Case's success; many believed that it was not only down to the man himself but also the people he employed, who were seen as the best of their generation. Case products became recognisable from the emblem of an eagle perched on a globe, which sent out the statement that Case wanted worldwide acceptance.

However, the first tractor that Case produced met with little success, and its successor came in weighing just under 14 tons, so was a big, heavy machine. Case tractors arrived in Great Britain in reasonable numbers as part of the Lend-Lease scheme in 1941; the first was probably the Model L, which took part in a major tractor trial in the 1930s and was found to have excellent fuel economy. Case replaced the L with the LA in 1940, and this model remained in production until 1952, powered by a petrol kerosene engine. During its production run Case introduced features such as lights, power take-off and electric start.

These tractors were brightly coloured in flambeau red. This was a relatively new colour for Case, which had painted its earlier tractors in a dreary grey colour. Using a bright new colour gave customers a renewed impetus to use the tractors. The Case Model D and the DEX were probably the most recognisable Case machines to be found around Great Britain during the Second World War, being popular on both sides of the Atlantic and selling more than 100,000. Many can still be found today in private collections in Great Britain and are very sought after by collectors.

Following the war Case became a big tractor producer, but few models found their way into Great Britain, and no particular model stands out as being popular. The company's financial position was not strong, having overspent in certain areas of the business, and it needed someone to provide cash. Luckily a company called Tenneco bought Case, and provided just the kind of ownership it needed to move forward. Now that Case was in a sound financial state it was able to expand further, not only in the tractor market but also with other agricultural machines.

In 1972 Case bought David Brown, thereby finding a way into the British market. David Brown was an extremely well-known tractor brand, so renaming existing models Case/David Brown gave the company instant machines to market in Great Britain. As an added bonus, owning David Brown gave Case a factory at Meltham Mills, Huddersfield, where models like the 1210 and the 1410 were produced in the David Brown colours of orchid white and power red. In Great Britain these tractors were adorned with the David Brown badge, while in America they were known as Case until 1985, when the parent company of Case Tenneco bought International Harvester.

International Harvester had been more popular over the years than Case; its range had been more comprehensive, with both small and large models being

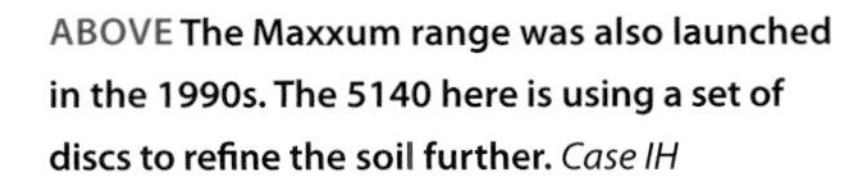

ABOVE **The Maxxum range was also launched in the 1990s. The 5140 here is using a set of discs to refine the soil further.** *Case IH*

RIGHT **Case Maxxum tractors have proved very popular with farmers in Great Britain. These early publicity photos used in some of the first brochures for this range show two Maxxum tractors ploughing and discing.** *Case IH*

LEFT **During the 1990s a new range of Case IH tractors was introduced, including the Magnum.** *Case IH*

ABOVE Tractors like these big Case IH examples produced today would have seemed impossible in the days of the company's humble beginnings. The 9390 and its modern day equivalent, the 450, are two of the biggest models made by Case IH.

well received. The take-over by Tenneco was perfectly timed as International was struggling financially. From now on the tractors were known as Case International; unfortunately this signalled the end for the David Brown name, and this once famous tractor manufacturer disappeared from the market. Not only did the name go but the factory at Meltham also closed and the workforce was made redundant. Meltham couldn't compete geographically with the International Harvester plant in Doncaster now owned by Case, which was ideally located for quick access to the motorway system; this was largely coincidence, as the factory had been there much longer than the motorway. By the end of the 1980s a new era for Case International was just beginning.

Case's Magnum range was to prove extremely popular with farmers; its styling was new and more rounded than the previous 'squared-off' tractors Case had offered. Soon Case Magnum models were seen on farms all over Great Britain where there was a need for a larger tractor. The range continues today and has evolved over almost 30 years, its universal appeal being helped by a number of features: a super-comfort well-laid-out cab, reliable performance, and bright LED lighting, which allows the driver to see easily around the tractor in the dark. For farmers needing a smaller tractor, Case also introduced the Maxxum range, which has a power output from around 100 to 141hp and today features surround-vision cabs providing excellent panoramic views that only let in 70dB of sound.

In 1999 Case merged with New Holland, owned by the motor car manufacturer Fiat, creating one of the biggest tractor brands to date. However, before the merger was allowed to go ahead Case had to shed a few assets, one of which was the Doncaster factory that International had set up in the 1950s. The other notable asset was the McCormick name, which had been retained since its last use in the 1970s; it and the Doncaster plant were sold to ARGO, an Italian tractor manufacturer that also owned Landini tractors.

Since the merger Case/New Holland has continued to be one of the main tractor brands in the world, and the ongoing production of the Maxxum and Magnum range has proven what a positive influence these models have been for the company. The introduction of the Puma range fitted perfectly within the existing models, and the company describes it as an efficient, multi-purpose tractor ideal for either field work or around livestock. Another notable addition to the Case line-up in recent times was the reintroduction of the Farmall name; as with the original Farmall tractors, these new machines are designed for multiple farm jobs.

From small beginnings started by one man to the major global brand that Case is today, no doubt Jerome Increase Case would be extremely proud of what has become of his company, and amazed at the amount of products carrying his name.

ABOVE **The Case Magnum and Maxxum CVX ranges continue the company's initial success.** *Peter Phillips*

Claas

The history of Claas stretches right back to 1913, the company named after its founder, August Claas. He was interested in the harvesting side of farming, and at first produced reapers, then successfully developing a knotter system that meant that straw could be packed and tied; this led to the design of a whole new machine that became what we know today as the baler. Not content with this, Claas was soon producing a combine harvester that was initially trailed behind a tractor and was powered by an engine attached to the front, as tractors of that period either had insufficient power to run them, or simply lacked a power take-off (PTO). Just after the Second World War Claas introduced a self-propelled combine harvester, a stand-alone machine that did not need a tractor to power it. Ever since, this well-known company from Harsewinkel in Germany has been famous for producing combine

RIGHT **Claas was already a well-established brand when it bought Renault to enter the tractor market. This early brochure from 2004 contains tractors originally designed by Renault.** *Claas UK*

BELOW **The first Claas tractors on the market looked very similar to Renault's models. This Ares 656 has the same name that Renault introduced.** *Claas UK*

ABOVE **Spot the difference! The Renault Ares was almost identical to the Claas Ares seen in the previous picture, the colour scheme being the only obvious difference.** *Renault Agriculture*

harvesters, and has also been a leader in developing forage harvesters, round balers and large square balers, so it seems hard to imagine that it took till 2003 for the company to enter the tractor market.

Buying a majority share in an existing tractor manufacturing company is one way of entering a highly competitive market, and acquiring Renault Agriculture allowed Claas to do just that. Renault only had a relatively small market share, which was mainly in France but also reached Great Britain. Claas was not interested in keeping the Renault name, so the tractor side of this famous car-maker disappeared. The excellent reputation of Claas in the harvesting sector allowed the company to quickly become a major force in the tractor market.

The first tractors produced when Claas took over were just rebadged Renault models, painted in the Claas colours. The Le Mans factory in France, set up in 1920 by Renault, is still the centre of tractor production, and constant updates to this facility have been key to Claas's success. Relentless development in areas such as a new cab assembly line and an environmentally friendly painting facility has seen Claas tractors become popular everywhere, including Great Britain.

In 2014 Claas celebrated 10 years of tractor production, announcing that it had made more than 100,000 tractors during that period. By 2018 the company aims to be producing and selling 25,000 tractors a year, and is well on target to achieve this milestone. Claas also has one of the most comprehensive ranges of tractors available in Great Britain, ranging from 72hp to a whopping 525hp. Its most popular tractor is the Arion range, which is available in 90-184hp; in 2016 Claas won the IF Design award for the tractor's panoramic cab, which features a convex windscreen with no frame between it and the sunroof, allowing an uninterrupted view when using a loader.

The Axion is the next step up, with power output of between 205 and 410hp from the six-cylinder Fiat engine. This ultra-modern engine is also extremely clean, meeting the stage IIIB (tier4i) emission requirements. This high-powered model in the Axion 900 series was introduced in 2011 and features CEBIS, a centre for the electronics all easily placed in the right armrest.

Having so comprehensive a range of models is what makes Claas tractors so popular, and the Xerion is the biggest. A centrally mounted cab and four equally sized wheels makes the Xerion unique in the Claas range. The design is very similar to the Mercedes-Benz MB-Trac, which was phased out in the 1990s; ironically the Xerion is powered by a Mercedes-Benz six-cylinder diesel engine with horsepower ranging from 419 to 520. Having an unparalleled view of the implements attached to the rear of the tractor from its cab is every driver's dream, so imagine the view you get from the Xerion's cab when, at the press of a button, it rotates through 180 degrees within 30 seconds so you can get a clear view of rear-attached implements.

LEFT This convex windscreen is of great benefit when using the tractor with a front-end loader. *Claas UK*

BELOW Claas soon grew the tractor side of the company, and in Great Britain they became very popular machines. Here we see a Claas Axion 340 mowing grass with a Khun mower conditioner. *Ben Phillips*

TOP RIGHT This impressive tractor is the Xerion, the biggest in the Claas range. *Claas UK*

BOTTOM RIGHT The Claas 900 Axion was introduced in 2011. *Claas UK*

CLAAS
5000
XERION
TRAC

CLAAS
CLAAS
930
AXION

David Brown

LEFT **David Brown's first tractor was produced in conjunction with Harry Ferguson and was called the Ferguson Brown.** *AGCO Archives*

Originally a gear system manufacturer set up in 1860 by David Brown himself, it took around 60 years for the company to enter tractor production. Sir David Brown, grandson of the founder, started building tractors in 1936 after a successful meeting with Irish inventor Harry Ferguson. These two men were very similar as they were great engineers who enjoyed solving problems. The Ferguson Brown was the first fruit of their partnership: Ferguson developed the design and Brown had the facility to build it, but the only name shown on the tractor was that of Ferguson. The hydraulic system made this little tractor unique at the time, but despite featuring this revolutionary invention the high price-tag led to slow sales. Each man blamed the other, and the cracks in their relationship began to show. David Brown had ideas on the way forward in tractor design and secretly began to build a new machine, while Harry Ferguson travelled to America to meet Henry Ford and discuss building a new tractor. In 1939, with the Second World War about to start, Brown and Ferguson's professional relationship ended and David Brown was ready to start production on a new tractor, while in the short term Ferguson was not.

The David Brown VAK1 was introduced with the three-point linkage that Ferguson had designed and which had been fitted to the Ferguson Brown; however, the VAK1's hydraulic system was sufficiently changed so as not to infringe Ferguson's patents, and his former partner could not complain. The bodywork was streamlined and featured a cowl surrounding the driver in an attempt to protect him from the elements. The design was carried over to the VAK1's replacement, the more famous Cropmaster, and later the Super Cropmaster. These tractors were also made into tugs for airfields, pulling not only planes but also the bombs to load into them prior to take-off. With sales reaching 60,000, David Brown was becoming a very rich individual. New, simpler and cheaper versions of the Cropmaster, the 25 and 30D, were introduced in 1953, but the Ferguson TE20 was dominating tractor sales; this was ironic, as it was designed and built by the man that David Brown had fallen out with over future tractor designs.

The 900 came onto the market in 1956; this 40hp model was a whole new tractor, but was to prove very problematic and gave the company a poor reputation. It was soon discovered that the fuel pump was to blame, which was not made by David Brown but by

RIGHT **The David Brown VAK1 started a successful run of tractors for this British company.** *Peter Phillips*

CAV. As these problems were affecting sales, the company brought its replacement, the 950, into production more quickly than had been anticipated. The Implematic version appeared a year later and, as with the 900, was offered with petrol TVO and diesel engine options; as with most tractors, the diesel was now the most popular. The PTO and hydraulics had a live system, meaning that the implement could continue to be powered by the engine even if the tractor was stationary; it also had a depth control via a top link sensing valve. A new idea called the Selectamatic was introduced a few years later, and David Brown claimed that it was the easiest hydraulic system anyone could wish to use; a single lever controlled everything. As with most things that

ABOVE By the time the Cropmaster came along David Brown tractors were becoming increasingly popular. *Jean & Malcolm Cooper*

LEFT Two people could ride on the Cropmaster's bench seat. Here they are using a potato spinner in the mid 1950s. *Jean & Malcolm Cooper*

BELOW A further advancement came when David Brown introduced the Super Cropmaster. This one belonged to Ken Briggs in Trimpley, where he is using a hay turner. *Ben Phillips*

LEFT **Cropmaster tractors were converted for all manner of different jobs outside agriculture; this one hauled trailers of bombs to aircraft for the RAF.** *Ben Phillips*

BELOW **New David Brown tractors were introduced in the late 1950s – this is the 950.** *Peter Phillips*

ABOVE **After years of producing tractors in red, David Brown changed to a white and brown colour scheme in 1965.** *Ben Phillips*

manufacturers claim to be simple, it turned out that the David Brown Selectamatic was more complicated to master.

The colour scheme of David Brown tractors over the years had gone through red, red and blue and red and cream, but by the time the 1200 came onto the market the white and chocolate brown colours had been used for a few years. This became the most familiar David Brown colour throughout the late 1960s. The 1200 replaced the 950 and after a few months in production the power was increased to 72hp, and a diff-lock and lights were also fitted as standard. A couple of years later a four-wheel-drive unit was added to the range, a system that was then becoming more in demand.

The American market, which was very lucrative for British companies, was proving good for these tractors proudly made in Great Britain, and 1971 saw a number of new David Brown tractors introduced onto the market: the 885, 995, 996, 1210 and 1212. These were now painted with white tinwork as before, but the dark chocolate brown was replaced with a bright orange. The company decided to expand the factory at Meltham Mills, Huddersfield, to cater for the demand across the Atlantic, but unfortunately it overstretched itself financially and in 1972 was forced to sell, American tractor company Case coming to the rescue.

The take-over by Case was initially a good thing, as development in areas such as Hydrashift transmissions on the 1212 received the attention they needed. In 1974 a new 1412 model joined the line-up, with a 91hp 3.6-litre four-cylinder turbocharged diesel engine. Cabs in the mid-1970s were undergoing a transformation due to new laws, and of course David Brown adhered to this. The 90 series featured a wide range of tractors from the 1190 model, of 48hp, right up to the 1690. This had the highest horsepower David Brown ever achieved in a tractor built on the Meltham Mills production line. The cabs on the 90 series often suffered from hot temperatures, but even with this problem for a time they topped tractor sales in Great Britain.

The Case name was an ever-increasing feature on David Brown models leaving Meltham Mills, so it was not surprising when the replacement for the 90 series, the 94, had the David Brown name replaced totally by Case. A name that had been around in agriculture since the beginning of the Second World War had now disappeared. The Meltham factory, which had made every David Brown tractor, also disappeared; Case had merged with International Harvester and decided it had no need for both factories, so the International facility in Doncaster was favoured for all tractor production from then on.

ABOVE The David Brown Selectamatic 880 went through a few styling and colour changes during its production. *Ben Phillips*

RIGHT By the time the 1210 was in production the colour had changed again, and the Case name was on the bonnet. *Ben Phillips*

Farmall

In 1924 most International Harvester tractors were big and cumbersome and could pull heavy implements. They could also work a drive belt from a pulley that was fitted by the tractor's brake or clutch pedal, which could run threshing boxes and similar implements. What was needed was a small but powerful tractor that could run a belt and pull implements, but was also light enough to work amongst the crops. The first Farmalls were row-crop tractors and were aimed at John Deere models of the same configuration, which generally consisted of two front wheels positioned together. International had seen how the Fordson tractors had been produced, so they began mass production of the Farmall tractors in a similar way.

The Raymond Loewy-designed Farmalls became among the most popular tractors in Great Britain, and were considered to be some of the freshest and best-designed vehicles of their time. One of the smallest models was the A; this had the steering wheel and seat offset, known as 'culti-vision', which gave the driver a clear view of the crops and any mid-mounted implements. Production of the Farmall A and B models started in 1939 and continued on well after the Second World War until 1947. It was the war that had prompted their export to Great Britain to help with the agricultural drive that was needed to help feed the nation, and Farmall benefited greatly from this new market opportunity. Also in 1939 two larger models, the Farmall H and M, completed the range. In Great Britain the H was probably more popular than the M, and sold extremely well with just under 400,000 produced between 1939 and 1953. A four-cylinder engine, six-speed transmission and a 6-volt starting system all contributed to the model's success.

If the Farmall H was one of biggest, the very smallest was the Cub, which was introduced in 1947 and was smaller even than the A. As with the A, the steering and seat were offset and the front axle was adjustable; the four-cylinder petrol engine was made by International Harvester. The original tractors had 'McCormick' on the bonnet, but this tractor was always affectionately known as the Farmall Cub.

The Farmall brand had been very good for International and had achieved high sales figures, at times outselling tractors from rivals John Deere and Ford. In Great Britain the Cub was really one of the last Farmall tractor to be sold in any great numbers, as in 1973 the name was dropped in America and all models now appeared under the International Harvester name.

ABOVE RIGHT **The Farmall A was one of the smallest tractors in the range.** *Ben Phillips*

RIGHT **Introduced in 1939, at the same time as the A, the Farmall M was one of the larger of the manufacturer's models.** *Ben Phillips*

LEFT Together with the H, the M was also at the higher end of the Farmall range. *Ben Phillips*

BELOW Normally Farmall tractors were red, but in 1953 a batch was painted gold to commemorate the Queen's Coronation, like the Super BMD pictured at a local show. *Peter Phillips*

RIGHT **The Super BMD featured a diesel engine and signalled that this was now the preferred choice of customers.** *Peter Phillips*

BELOW **The Farmall name disappeared for decades, but recently Case IH has brought it back in the current range of tractors. This Farmall C Series model is seen here with a hay rake.** *Case IH*

Fendt

In the early 1930s the first Fendt tractor, a 6hp Dieselross, was built by the Fendt brothers, and was a basic machine rather like an engine bolted to an iron frame. During the Second World War the factory built engines that would run on any flammable liquid, giving this company a good reputation of being able to adapt engines in times of need. After the war production of the Dieselross tractors resumed, and in 1958 a tool carrier joined the line-up; this 12hp machine was well received in its native Germany, winning a top agricultural award for its design. During the 1950s and 1960s Fendt brought out tractors with the names Favorit and Farmer, and these would become well known in later decades in Great Britain.

LEFT **A brochure for Fendt Farmer models from the 1980s and 1990s.**

In the early 1970s Fendt considered entering the British market and decided to appear at the country's main tractor show. The timing could not have been better, as there was not a great deal new in the tractor world and this brand received all the attention. New dealers soon came forward and Fendt was well on its way to becoming an established tractor company in Great Britain. The one snag, however, was that they were expensive compared to every other tractor on the market, although the quality was clear to see and many farmers still seemed interested, despite the price-tag.

The first models introduced into Great Britain were the updated Favorit and Farmer models. Working in their favour was the fact that they were available without a waiting list, a problem that was plaguing other brands at the time. Some were unable to supply a tractor for months, which meant that if a customer needed a new one at harvest time, he might be told that it would not be ready till Christmas – being able to purchase one there and then was what customers wanted to hear. Fendt had adopted a colour scheme of green and red from the beginning, and the models produced in the 1970s and 1980s carried on this tradition. The tractors had a distinctive long, square bonnet, which generally housed a lovely sounding six-cylinder engine.

Fendt developed a transmission system called Turbomatik, which was revolutionary at the time it was designed and worked using fluids instead of the more common mechanical system found on other tractors. Drivers reported that this was particularly useful when towing heavy objects and the engine revs dropped; unlike their mechanical counterparts, they wouldn't stall or slip. During the 1980s Fendt was seen as the top end of quality and at the forefront of technology with digital displays, automatic diff-locks and electronic hydraulic linkage, but was still finding its feet in the British market. Big changes were to come in the 1990s, when Fendt appointed new distributors in Great Britain and introduced a fresh new design of bonnet, leading to considerably increased sales. But this upsurge was not due solely to new importers or new styling; instead Fendt had relented in its strict pricing scheme, and now discounts were available due to significant competition from other well-established brands in Great Britain.

BELOW **The Fendt 250s models proved popular in Great Britain as they were compact, but still could manage jobs like ploughing.**

ABOVE The long, square bonnet gave Fendt tractors a distinctive look. This Fendt Favorit 926 is towing a Claas grain trailer.
AGCO Archives

LEFT Fendt had a deep green and red colour scheme that has been largely the same throughout the decades that the tractors have been found in Great Britain. This is a Farmer 308C model.
AGCO Archives

ABOVE The Vario system in modern Fendt tractors has proved extremely good. This 818 Vario is riding alongside a Fendt Katana while it harvests maize in Trimpley, Worcestershire.
Ben Phillips

RIGHT The Fendt Favorit Turbomatik was a common sight on many British farms.
AGCO Archives

Matters improved even more in 1994 when the highly anticipated Xylon was introduced, which was different from anything the company was producing at that time. This new machine featured a centrally mounted cab and drew upon the success of the Mercedes-Benz MB-Trac; it was received to great acclaim and really started to put Fendt on the agricultural map. The company was already known for its Turbomatik transmission, so when it introduced its Vario system it pushed tractor technology just a little further.

The biggest news, however, came in 1997 when it was announced that AGCO, one of the biggest tractor manufacturers in the world, had bought this German company. Fendt had already been around for more than 60 years, and with its new owners it could continue to be at the forefront of tractor technology. The company was certainly attractive to AGCO, with its technological innovation and its Vario transmission system, which had proved a good step forward from the Turbomatik. Fendt's Xylon system tractors had sold well and AGCO wanted this brand as part of its ever-increasing portfolio. At the turn of the new century, under AGCO ownership Fendt expanded its tractor range and even started building combine harvesters, together with round and square balers. Today there are tractors ranging from 70 to 517hp and nearly all feature the Vario transmission, which has allowed Fendt to mix with the big boys for so long that it is now one of them. It is also nice to see the colour scheme of green and red still being used, which makes the tractors instantly recognisable as Fendt products.

RIGHT Having AGCO behind it has allowed Fendt to carry on being a premier brand today – this is one of its latest models, a 936 Vario. *AGCO Archives*

BELOW Fendt is renowned for producing big tractors capable of a wide range of tasks. This dual-wheeled Favorit 926 is pulling a Lemken plough. *AGCO Archives*

Ferguson

One man revolutionised farming in the 1940s, and his name was Harry Ferguson. This Irish engineer had an interest in aviation, motorcycles and agriculture, but it was for tractor design that he was to become most famous. Born in Dromore, County Down, in 1884, it did not take him long to become interested in planes and motorcycle racing, and in an era when aviation was in its early form he visited many air shows and even built his own plane with the help of his brother Joe. When he took his first flight in 1909 he was the first Irishman to do so.

By 1911 he had fallen out with his brother regarding aircraft and the obvious safety concerns at a time when aviation crashes were all too common. When he started selling Overtime tractors he began to understand the problems involved with pulling implements behind; tractors of the time only had a drawbar and, if an implement got caught by an obstruction, the front of the tractor would probably rear up and more than likely hurt the driver in the process. Ferguson started to develop a method of making implements fit the tractor in such a way that they became part of the structure, and for this he used a few of Henry Ford's machines to try out his ideas. Settling on a mechanical three-point linkage system, he started demonstrating this and soon Henry Ford expressed an interest, saying that he could incorporate the idea in the Ford/Fordson tractors in his factory. However, Harry Ferguson was a tough businessman and the two failed to agree.

ABOVE **Harry Ferguson, the Irish engineer who revolutionised farming.** *AGCO Archive*

BELOW **The Ferguson Brown featured the famous three-point linkage, made by David Brown and marketed by Harry Ferguson. This one is pulling some scuffles to break up the soil.** *AGCO Archive*

Developing the system further, and now using hydraulics, it was even better than before, but he still needed someone with a factory to actually fit it to a production model. After a few problems he managed to strike an agreement with David Brown, and the Ferguson Brown tractor was born. This simple little machine had a hydraulic lift that Harry Ferguson had spent many years developing. The Ferguson Brown began with a four-cylinder 20hp Coventry Climax engine, then later versions had a David Brown-built example, based on the Climax design. At first the Ferguson tractor did not sell very well, largely due to the price being just over £220; at that time the economy was not strong, and many farmers bought the cheaper

ABOVE LEFT **The Ferguson Brown did not sell well and led to Harry Ferguson and David Brown going their separate ways.** *Ben Phillips*

ABOVE RIGHT **The Ferguson TE20 was to become one of the most iconic tractors ever produced.** *Peter Phillips*

Fordson. This lack of sales led to friction between Harry Ferguson and David Brown, and their relationship started to deteriorate.

In October 1938 Ferguson went to see Henry Ford again, taking with him a Ferguson Brown tractor to demonstrate. Ford liked what he saw, and this time a successful deal was done, becoming famously known as the 'handshake agreement'. In 1939 a Ford Ferguson 9N rolled down the production line and featured Harry Ferguson's hydraulic system. This new Ford tractor was far more modern than the Brown, and both parties were happy as they developed further models together. However, this agreement that had started so well finally ended in a bitter divorce in 1947 when Henry Ford's grandson, also named Henry Ford, pulled out of it. Ford was still using Ferguson's designs while the patents were still in force, and this prompted Harry Ferguson to take legal action. The court case involved not only a lot of money but also a lot of stress, and when it was resolved Harry Ferguson had won a $9 million settlement.

While the court case was in progress Ferguson was without a tractor in production, so he approached Sir John Black, who owned a factory in the Midlands that produced Standard motor cars and was well placed in the centre of Great Britain in Banner Lane, Coventry. In 1946 the first Ferguson TE20 rolled off the production line at Banner Lane. Some of the early models were either fitted with the Standard Motor Company petrol engine or the Continental petrol engine; the Standard factory was unable to satisfy Ferguson's demand for these engines, so the Continental was imported from the USA.

As well as being a great engineer, Harry Ferguson was also great at marketing his TE20 tractor; at any opportunity he would demonstrate the versatility of the machine and the added benefits of owning one. Pictures of him using a Ferguson TE20 on his large country estate at Abbotts Wood in the Cotswolds showed how capable this tractor was. The 'grey Fergie', as it became known, was a great success, although some farmers were not so enthusiastic about it and its new system; after years of owning a tractor that simply dragged an implement, they would now have to buy a whole host of new attachments to get the most from their new Ferguson.

As production continued most farmers came to love this little tractor, and many variations were developed such as narrow, industrial and vineyard. A new petrol TVO version joined the already straight petrol model, providing greater fuel economy. Harry Ferguson was initially against diesel engines being fitted as he thought they were dirty, but eventually a diesel joined the range.

Probably the Ferguson TE20's greatest triumph was not in the green fields of Great Britain but in the frozen white land of the Antarctic. In the late 1950s four petrol and three diesel TE20s set off with Edmund Hillary to the South Pole, some fitted with tracks and front skis while others had an extra wheel on either side. How these tractors faired in this harsh environment was testament to how good they were – coping with -23 degrees and slippery ice is no easy feat, but the TE20 never let the expedition down. When they got to the South Pole the Ferguson TE20 was the first vehicle to be driven to one of the most inhospitable places on earth.

As the TE20s were conquering the Antarctic, a new model was introduced, the FE35. While the TE20 was all grey in colour, the FE35 had grey tinwork and gold castings, and was sometimes known as a 'gold belly', or simply a 'grey and gold Fergie'. This new tractor had a raft of updated features on the hydraulics, dash and seating, not to mention new styling and engines. However, the diesel engine soon gained a reputation for bad starting and many owners got very annoyed when their tractors simply wouldn't start in certain conditions. Luckily, within a few years these were rebranded Massey Ferguson and a new Perkins engine came to the rescue.

ABOVE Ploughing was the Ferguson TE20's forte, as shown in the Ferguson System logo. The logo was a side profile of the TE20 with a two-furrow plough. *AGCO Archives*

RIGHT Harry Ferguson would demonstrate the capabilities of this little tractor at every opportunity, including here at his home at Abbotts Wood in the Cotswolds, where he is dredging weeds from a water course. *AGCO Archives*

ABOVE A saw bench was just one of the attachments made for the Ferguson TE20. *AGCO Archives*

BELOW The new Ferguson FE35 fitted with the 23c diesel soon gained a bad reputation for poor starting. The tractor in this photograph is in on test as shown by the white number 21 on the bonnet. *AGCO Archives*

Even when the merger with Massey went through, Harry Ferguson stayed on in a managerial capacity, but in true fashion he soon disagreed with other management figures. This time it was over a new model, the Ferguson LTX, which was a bigger tractor that Ferguson felt the market needed. Even though prototypes were built, Massey Harris decided against producing the model. Ferguson was furious and nearly everything to do with the project was destroyed; he promptly resigned before production of the TE20 had finished. This great engineer was suffering with depression, had trouble sleeping and was hard of hearing, but still managed to fill his time working on other projects. He was becoming increasingly interested in the use of four-wheel-drive systems, sometimes outside agriculture altogether. In 1960 Stirling Moss took the chequered flag at the Oulton Park Gold Cup in a Formula 1 car entered by Rob Walker. This included a four-wheel-drive system designed by Harry Ferguson, and a similar system would in the future be fitted to a production road car.

ABOVE **This Ferguson FE35 in grey and gold is fitted with a banana-loader and a muck-spreader, two lovely implements made for the model.** *AGCO Archives*

One evening in October that year, Ferguson retired to have a bath at his Abbotts Wood estate, and was never seen alive again. It was rumoured that his death was due to a barbiturates overdose, but to this day no one really knows whether it was accidental or not.

Harry Ferguson's legacy continues to this day, as nearly every tractor produced is fitted with a three-point hydraulic linkage. The little grey tractor he designed and built became a legend in its own lifetime and today is sought after by many collectors across the world.

Fiat

This famous Italian brand is more synonymous with producing motor cars, but it has also been building tractors since 1918. The first model produced was the 702, which soon became extremely popular throughout Italy, selling a couple of thousand. In fact, around the same time tractors were leaving the Fiat production line in large numbers each year, which also pleased the Italian Government, which favoured home-made over imported products. Agriculture in Italy was strong, the climate was kind and Italians love fresh healthy produce, so on countryside farms on steep terrain a tractor was a very useful commodity. The Fiat 702 was exported to other European countries including Great Britain and, as with most continental tractors, it was well made with good chunky metal. However, being built so well the Fiat was expensive at just under £600, when rival tractors were considerably cheaper. During the Second World War the Fiat tractor factory was commandeered by the Germans to repair vehicles used by the army; tractors were still made, but in those tight financial times people were not looking to buy them.

After the war the big, heavy Fiat was not what farmers wanted, so the company needed a solution. It did not need to look far – in fact, only to the motor car side of its business, where it produced the Topolino, a small popular car that provided the inspiration to

ABOVE **Fiat were making heavy tractors until they saw the popularity of their small Fiat Topolino car. So in the 1950s they built a small tractor – the 211Rb is shown here – known as La Piccola.** *CNH*

BELOW **As the 1980s dawned the need for bigger tractors allowed Fiat to return to what they originally started off doing, making large models such as this 110-90. This vehicle is demonstrating its capabilities by lifting a Lely cultivator and Hassia seed drill.** *CNH*

replicate it for the agricultural industry. The tractor produced was the La Piccola, a machine that featured a distinctive long bonnet but, more importantly, was light and nimble and had good manoeuvrability, just right for Italian vineyards. Its colour scheme of bright orange tinwork and dark brown castings was to adorn Fiat tractors for decades to come. The model proved a good move and sold well, and within two years around 20,000 had been sold, which established Fiat as a major player in the agricultural machinery sector. The 1960s and especially the 1970s were good times for tractor production; technological advancement, increased safety and the need for good powerful tractors due to increased field sizes allowed Fiat to go back to producing bigger and heavier tractors.

By the 1980s Fiat was one of the biggest manufacturers of tractors in the world, and as with most companies connected to the Agnelli family this was no surprise. The company started to produce

BELOW **At the start of the 1990s it was time for a fresh design, as shown here in the Fiat F140 which was powered by a 5.9-litre Iveco six-cylinder diesel engine.** *CNH*

ABOVE **The 90 series was a popular set of tractors during the 1980s. Their distinctive cabs designed by famous Italian designers Pininfarina helped them to stand out from the crowd.** *CNH*

tractors that were styled by legendary Italian designer Pininfarina, which was also responsible for a great number of Ferrari cars, in which Fiat owned a part share. Fiat tractors were renamed Fiatagri in the mid-1980s, and by now the bright orange had been replaced with a much darker shade. The super-comfort cabs that were developed to give the driver as much comfort as possible had noise levels at only 80dB, air-conditioning was available as an option, and the cabs had a distinctive diagonal line at the back of the door. The 1980s was an important time for tractor electronics, and Fiatagri had its own known as Agritronic, which was an onboard computer system that could be installed in the 'Big Fiats', as they were known. The engines in these high-end tractors were of 5.8 or 8.1 litres, giving a power output of 140 and 180hp, and together with these powerful engines a four-wheel-drive system was also fitted. The need for four-wheel-drive tractors at this time was ever-increasing, and Fiat had developed this idea many years previously; indeed, the Italians really pioneered four-wheel-drive tractors.

This style of Fiat tractor, with the long square bonnet and the famous diagonal door frame, was seen on a number of farms in Great Britain, their presence being particularly strong in some areas. During the 1980s Fiatagri produced its 500,000th machine, in a range that included crawlers and tractors. As the 1990s dawned the chance to buy Ford New Holland was too good an opportunity to miss; Ford was making good tractors and, since it owned New Holland, the implements also had a good reputation. Obviously the Ford name needed to be dropped, and although Fiatagri had been around for many years, New Holland was a better-known name in agricultural circles. In certain European markets the existing Ford models were rebadged and recoloured in the familiar terracotta red of Fiat tractors, while in Great Britain they carried on being blue and black and were branded New Holland. The names Fiat or Fiatagri disappeared in Great Britain, and New Holland became known as a tractor manufacturer instead of just a company that made implements.

With Fiat in full control, the brand was now stronger than ever and had a massive market share, which was further increased when

RIGHT The smaller Fiat 88-94 tractor fitted with 3.9-litre Fiat engine was still capable of using a four-furrow reversible plough. This model was in production from 1993 to 1996. *CNH*

BELOW When Fiat took over Ford/New Holland the tractors were rebadged as New Holland. This terracotta-coloured M160 is almost identical to the blue New Holland tractors found in Great Britain. *CNH*

it merged with the American tractor giant Case to create Case New Holland (CNH). To further emphasise how big this company had become, before the merger was allowed to take place Case had to dispense with certain assets, one of which was McCormick. Today, although the Fiat name does not appear on any tractors sold in Great Britain, it does own one of the biggest tractor companies in the world.

Ford

Henry Ford had pulled out of tractor production in the USA in 1928, but a decade later was contemplating a return in some way. The Ford Company of Great Britain continued to make Fordson tractors in Ireland and Great Britain, which were exported back across the Atlantic. Henry Ford wanted something more special than the Model N, and it took a meeting with Harry Ferguson to persuade him to start producing tractors again. Irish engineer Ferguson knew the Fordson tractor well, as he had fitted an experimental lift system to an F to demonstrate it. Henry Ford saw the advantages, and in 1939 they struck a deal; the first tractor produced by the partnership was the Ford 9N. This was fresh-looking, featured a hydraulic lift system designed by Ferguson and was built at the Ford factory in Dearborn, Michigan. There were plans for it to be built in Great Britain at the Dagenham factory, but Henry Ford did not have full control of this facility, so that never materialised. The 9N was replaced in 1942 by the 2N, partly due to wartime shortages of certain parts, and partly because by law price increases could only made on new models.

Henry Ford II took control from his grandfather in 1945 and, as with most new management, changes were not far behind. These changes were greatly needed as, even though the new Ford tractors were selling well, the company was losing money fast. The deal with Harry Ferguson was proving fruitful for him, but not for Ford, so Henry Ford II cancelled their agreement. This resulted in a bitter court case between Ford and Ferguson; Ferguson triumphed in the end, though it took its toll on both parties.

The 8N was the final model in this series of tractors and became the best-selling tractor ever in the USA. Since it came out after the court case, Ford didn't want to contravene the patented Ferguson hydraulic system, so made certain changes to it. The colour scheme was also changed from the 9N and 2N, which were all-over grey, to a very light grey tinwork and red castings for the 8N. A new four-speed gearbox and an increase in power were enough to improve this tractor over its predecessors.

Until 1964 there were Fordson tractors built in Great Britain that were exported to America and Ford tractors built in America that were exported to Great Britain. Both companies had remained separate over the years but were in constant talks with each other. In the early 1960s it was decided that all tractors, whether built in Great Britain or America, would be called Ford. The first Ford tractor built in Great Britain was the 1000 range, which arrived in 1964 and was received with great fanfare. Tractor production moved from Dagenham to a new factory in Basildon, where this newly designed tractor rolled off the assembly line.

To ease in the new 1000 range, Ford labelled the early tractors by their old Fordson names. The 2000 Dexta was the smallest, and the most popular engine was the three-cylinder diesel, which produced 37hp. The next model was the 3000 Super Dexta, and again the diesel was the most popular, producing 47hp. As the range expanded, so the tractors got bigger. The 4000 Major was quite big for

ABOVE LEFT **The Ford N series featured the Ferguson system designed by Harry Ferguson. He needed a tractor for his invention so Henry Ford produced these tractors to do this.** *Ben Phillips*

LEFT **The Ford 8N had a different colour scheme from the previous N series models.** *Peter Phillips*

ABOVE Ford opened another British factory in Basildon in the early 1960s to produce a whole new range of tractors. *New Holland*

its day and, having 55hp, was a very useful tractor. The biggest, giving 65hp, was the 5000 Super Major.

Ford had developed a revolutionary new gearbox, christened Select-O-Speed, which was an optional extra. The gear stick was replaced by a lever that could be shifted on the move; it was a great idea, but tended to have a mind of its own and collisions could result from selecting the wrong gear! Some say it was ahead of its time and, had it been more refined, it might have been better received by the farming community; instead, some of those who had bought into this idea had the gearbox replaced with a conventional one. Some gave it the nickname of 'Jerk-O-Matic' due to its constant jerking while shifting between gears. At the end of the 1960s Ford updated the 1000 range and dropped the old names; now they were simply known by their numbers.

In 1975 Ford produced some beautiful tractors as the latest models in the original 1000 range, which today are highly collectable classics. They included the 2600, 3600, 4600 and 7600, all popular models in Great Britain, and Ford was now becoming a major player in the tractor market. The 700 series followed shortly, with an updated shape as well as a better working environment for the driver. As with most other tractor manufacturers, Ford responded to new Government laws by making tractor cabs quieter and safer, its answer being the Q cab.

As the 1980s dawned a new 10 series was brought onto the market, fitted with three-, four- and six-cylinder engines producing 47 to 110hp. Ford offered synchronised transmission on all the three-cylinder models, which offered rapid gear shifting under light loads. With the four- and six-cylinder models, constant mesh or synchro shift transmissions were available. The 1970s Q cab was updated and was now called the Super Q, even quieter and with more driver comfort. A new seat boasted seven comfort positions, extra padding, armrests and shock absorbers. Throughout the 1980s the Ford 10 series could be found on farms everywhere in Great Britain.

To serve farmers needing a bigger tractor, Ford introduced the TW series in the 1980s, with six cylinders producing 136-186hp. The air-to-air intercooled and turbocharged engine fitted in the range-topping TW35 was unique in tractor production. The TW series also

FORD
5000

LEFT A new breed of Ford tractors soon rolled off the production line at Basildon. This shows the Ford 5000 being driven off the line. *New Holland*

ABOVE Ford tractors waiting to be shipped from Dagenham in the 1960s around the world. *New Holland*

had the Super Q cab and benefited from a myriad of great features. On top of the cab there were six bright halogen lights that flooded the front and rear working areas in dark conditions. A four-wheel-drive system featured a centrally mounted differential, which improved the tractor's turning circle.

Around 1985 Ford bought New Holland, a firm famous for producing implements. Ford had seen its rival John Deere produce both tractors and implements, and wanted to do the same. At the end of the decade Ford celebrated 25 years of the Basildon factory by producing a Silver Jubilee special edition of the 7810, a limited run of tractors that had silver tinwork and cab and blue castings. The 7810 was a well-respected tractor in 1989, but many customers turned up their nose at a silver Ford. Farmers and contractors who owned a fleet of the same make usually liked them in one uniform colour, so having one silver tractor in a group of blue ones didn't sit well with some. This situation prompted a few dealers to have the Silver Jubilee 7810s sprayed blue. After an unpopular start, the

ABOVE **The Ford 3000, fitted with Select-O-Speed, soon became a disliked addition and was later dropped.** *Ben Phillips*

TOP RIGHT **At the beginning of the 1970s the 1000 range was updated, the most noticeable change being the styling.** *Ben Phillips*

BOTTOM RIGHT **The whole updated range, including the Ford 5000, had a completely new appearance.** *Ben Phillips*

FORD 3000
FORD

FORD 5000
FORD

FORD
FORD
4000

surviving Silver Jubilee models have sky-rocketed in value, and today it is a much sought-after tractor with collectors.

Next a very useful small/medium utility series was offered in the form of the 3930 and 4630 models. A new hydraulic system allowed these tractors to lift more than they appeared to be able to. A 55-degree steering turn gave it excellent manoeuvrability, and could be done with ease thanks to being hydrostatic, which meant that farmers found it useful around tight farm buildings. The cab was all new with two opening doors that led to a flat-floor cab, which had low noise levels. A dual-power synchro-shuttle transmission gave the driver an easy time and improved the tractor's efficiency. Fitted with grass tyres, these models were particularly useful for mowing parks and playing fields.

ABOVE **The Ford 7600 was just one of many popular Ford tractors found in Great Britain.** *New Holland*

LEFT **One of the biggest in the 1000 range was the 4000.** *Ben Phillips*

FORD
FORD

The 10 series had proved extremely popular for Ford, and sales had been strong not only in Great Britain but all over Europe. As the 1990s began the 40 series was due for release, which was going to have to be something special to top the 10 series. Ford had newly designed pretty much everything on the 40 series, although some of the parts were considered too good to redesign, so were left alone and fitted to the new model. The range started with the 5640, a four-cylinder 75hp tractor, and the 8340 six-cylinder 120hp completed the line-up. The first models off the production line had white cab roofs, but in 1995 these were changed to blue. As with most tractors around this time, the 40 series relied on electronic systems, although all but two of the models could be supplied with either electronic or manual hydraulic lifts.

Some drivers noted that the gearbox was not very smooth and tended to jerk; they also found that if the electrical solenoid that worked it needed to be replaced, the gearbox would have to be opened up in order to locate it.

LEFT **A typical sight on a British farm in the late 1970s: a number of blue Ford tractors doing all manner of jobs.** *New Holland*

RIGHT **The TW series catered for farmers who needed a bigger tractor.** *New Holland*

BELOW **The 10 series like this 8210 was introduced in 1982 with a raft of new features.** *Jane Brooks*

The Ford and New Holland merger was becoming even stronger midway through the 1990s, and the 40 series underwent more changes as a result. A Fiat-designed front axle was adopted, as was an updated interior. Another change was that the tinwork was being phased out and replaced with hard thick plastic.

Even though Ford's tractors had been popular and its motor cars even more so, hard economic times were affecting many businesses; for many years Ford had been looking for a way out of the agricultural market, and needed little persuasion to sell. So when Fiatagri expressed an interest to buy the whole business, a deal was soon done, and the name Ford on the side of the bonnet was quietly replaced with New Holland. From now on no tractors would feature the famous Ford name, and the surname of the man that had done so much in tractor production had gone, replaced with a name more synonymous with agricultural implements.

LEFT In 1989 a number of specially painted Ford tractors were produced to mark 25 years of tractor production at Basildon. These special edition tractors failed to sell as well as it had been hoped, due to them not being blue and expensive. Today they are sought after by collectors. *New Holland*

BELOW Ford updated the 10 series, which was becoming a well-loved range of tractors throughout Great Britain. As the company now owned implement firm New Holland, a lot of its publicity photographs included implements like this New Holland round baler. *New Holland*

Ford introduced a useful range of smaller tractors, exemplified by the 4630. Again a New Holland round baler is also featured. *New Holland*

ABOVE **British farmers couldn't wait to get their hands on the new Ford 40 series. This one is using a hedge cutter flail.** *Jane Brooks*

FORD

Series 40 75-120 hp Tractors
Models 5640, 6640, 7740, 7840, 8240 and 8340

FORD
NEW HOLLAND

All-new power and features designed to help you work more productively.

ABOVE **To great fanfare a new range known as the 40 Series was introduced in the 1990s.** *New Holland*

RIGHT **The 8210 produced around 100hp from its six-cylinder diesel engine.** *New Holland*

Fordson

Henry Ford will always be remembered for making things easier, whether that involved getting people from A to B more quickly and easily with a Model T car, or making farm work easier. His relentless quest to do this can be traced back to his childhood. Growing up just outside Detroit, Michigan, on a family-run farm, he saw just how tough life was for these workers. Heavy labour done by both humans and horses was evident every day, and as the 20th century dawned the young Henry Ford was well on the way to developing a tractor from automobile parts, and in 1907 the first tractor was unveiled.

The first Fordson tractors in Great Britain were imported from the USA around 1917, in order to boost British agriculture as the First World War was sapping all things mechanical. As with the Ford Model T car, the Fordson tractor was popular, being affordable and simple; the philosophy was that having such a machine available at the right price would spike people's imagination.

Having exported thousands of tractors to Great Britain, Henry Ford decided to open a factory on this side of the Atlantic, and settled on Cork in Ireland as the place to do this, wanting to develop employment in that part of the country. Opening after the war, the Cork factory would become synonymous with Fordson products. The first tractors built were 20hp four-cylinder machines that would run on petrol, kerosene and even alcohol! This tractor was known as the Model F. The radiator and fuel tank were all that covered the engine, and metal-rimmed wheels without any mudguards for protection made up this simple tractor. There was nothing unique about the tractor, but it was cheap and there was a dealer close to the people who needed it, so it proved extremely popular.

With the advent of trouble in Ireland in the early 1920s, the factory at Cork closed and Henry Ford moved production back to the USA. However, six years later, in 1928, after producing more than 650,000 tractors, Henry Ford stopped production in the USA. It is thought that rival tractor manufacturers had prompted Ford to do this, as he believed that their involvement in tractor production had taken away what had made the Fordson so special to his customers. However, Ford in Great Britain planned to restart production at the Cork factory, and the tooling was shipped back from the USA. Once in full production, Fordson tractors were now shipped back across the Atlantic. In 1930 the factory at Dagenham in Essex was opened, and all tractors were now built there. Cork closed again, but for good this time.

By now the Model N had replaced the F. This updated tractor was very similar in appearance, but had an extra 7hp. New rear wings gave protection from the rear wheels, the magneto electrical system featured a higher-voltage output, and the new rubber tyres gave better comfort instead of those hard metal rims. At the time no one really knew how important this tractor would be during the Second World War; the country needed feeding, which meant that

TOP **The Fordson F came to Great Britain around 1917. This basic tractor was started by turning the handle.** *Ben Phillips*

ABOVE **The Fordson N replaced the F in 1929 and was painted in a tan/orange colour until the 1940s, when the colour was changed to green.** *Peter Phillips*

ABOVE **A number of variations of the N were made, including this industrial model. They normally had a high top gear so they could travel faster. The RAF used this variant.** *Ben Phillips*

BELOW **To help camouflage Fordson tractors working in fields during the Second World War, they were painted dark green. Here it is seen with a trailed plough, which is just one of the implements they were famous for pulling.** *Ben Phillips*

many acres of farmland that had been left wild needed to be ploughed up. With many men away at war, male tractor drivers were in short supply, so the Women's Land Army was pressed into action. Women all over the country soon learned how to drive the Fordson N tractor. The first examples were painted blue and orange, then just orange; however, with the onset of war it was decided that such a bright colour would be too visible to German bombers, so a new dark green was adopted to help camouflage them.

In 1945 a new Fordson was introduced. The E27N was based around the Model N, but was bigger and taller and the blue and orange colour scheme returned now the war was over. The engine and transmission was carried over from the N, but the casting for the rear end was redesigned and now a hydraulic lift could be fitted. Customers buying this new tractor also had the option of full electrics and a front axle with a changeable width. As the E27N was bigger than the Model N, the latter's petrol TVO engine with which it was fitted was now struggling for power, so a factory-fitted 45hp Perkins P6 was offered, which transformed the tractor.

Fordson now wanted to bring out a whole new tractor, but the country was so short of essential supplies, having been brought almost to its knees by the war, that a new model had to wait till 1953. It was known as the E1a Major, with a new colour called empire blue, a lighter shade of blue than earlier Fordsons. More and more tractors were now using diesel power, so most Majors had a diesel engine, freshly developed at the Dagenham factory,

RIGHT The Fordson E27N was bigger than the N. This heavier tractor was useful for ploughing, as shown here in the picture.
Ben Phillips

BELOW The new E1a Major soon proved popular throughout Great Britain, seen here loading sheaves on a trailer fresh from a binder.
Peter Phillips

although petrol TVO versions were also available. Previous Fordson tractors had little in the way of tinwork and the radiator and fuel tank were exposed, but when the E1a was introduced it featured fresh styling that covered over most of what had previously been on show. By the late 1950s an updated version of the Major was on sale; this had a double clutch, a better gearbox and the engine now produced 51hp. Otherwise its styling was largely the same, and it was called the Power Major. Finally in 1960 the Super Major came onto the market, and was the last version of this popular range of tractors. Before production ended, the colour scheme had changed from blue and orange to blue and grey.

Mirroring the Major range, the smaller Dexta was an extremely popular tractor. The first models were quite similar in shape to the Major, only smaller. Fitted with a Perkins three-cylinder diesel or a Standard petrol engine, this tractor was very capable around the farm and could do most jobs; in fact, some farms in the late 1950s would only need a Dexta-sized tractor. In 1962 the Super Dexta replaced the first-generation Dexta, and slight styling changes, mainly to the front cowl, allowed this model to remain fresh. As with the Super Major, the Super Dexta was coloured blue and grey, the colour scheme signalling a new era when the Fordson name was set to disappear from tractors for good. The new models introduced in 1965 were to be called simply Ford.

TOP The E27N was replaced by a new model, the Major, during the 1950s. *Ben Phillips*

ABOVE The Super Major saw a colour change as well as other differences, this one in a private collection belonging to Alan Braithewaite. *Ben Phillips*

LEFT The Power Major was introduced in the late 1950s, as seen here in Salcombe, Devon. *Ben Phillips*

TOP RIGHT Many variations of the Major were introduced, including this Roadless version. *Peter Phillips*

BOTTOM RIGHT Another famous Major conversion was the Doe, consisting of two Majors bolted together. This was done to provide a more powerful tractor. *Peter Phillips*

221
AMR 936B

30 NWC

ABOVE LEFT **The Major's little brother was the Dexta.** *Jean & Malcolm Cooper*

ABOVE RIGHT **The Super Dexta was the final incarnation of this model.** *Ben Phillips*

BELOW **The Dexta was in direct competition with the Massey Ferguson 35. Here it is powering a binder for cutting corn which was the forerunner to the combine harvester.** *Ben Phillips*

International Harvester

As with most American agricultural companies of the age, this one was started by one man – Cyrus Hall McCormick. Way back in 1831 in Virginia this inventor demonstrated his reaper, which was horse-drawn; a few years later it had patent protection. A move to Chicago in 1847 saw him team up with his brother Leander and the company really took off. Success came quickly, partly due to the McCormicks' business acumen and also due to the recent railroads, allowing products built in Chicago to be transported with ease throughout America. Even in those days McCormick had a number of people nationwide who demonstrated his machines to potential customers.

By the start of the 1900s Cyrus McCormick Jr was in charge. The founder's son, he had taken over the business in 1885 and, merging with a handful of other companies, International Harvester was born. The Titan 10-20, made at the company's Milwaukee factory, was among the first International tractors to be found in Great Britain, and both there and in America they were competing with Fordsons. The International 8-16 Junior had the more familiar tractor shape that we know today, and a lot of the parts used in it came from the truck side of the company. Around 2,500 International 8-16 Juniors were exported to Great Britain. International Harvester's tractors were quite big and heavy, and their main rival, the Fordson F, was smaller and lighter; the Farmall was therefore introduced, which was a lighter and more

TOP RIGHT **The International Titan was introduced into Great Britain during the First World War.** *Ben Phillips*

CENTRE RIGHT **1918 saw the introduction of the International 8-16.** *Peter Phillips*

RIGHT **The later 8-16 Juniors were painted in light green.** *Peter Phillips*

FDD 55

INTERNATIONAL

TOP LEFT **The International 10-20 was introduced in 1922.** *Peter Phillips*

BOTTOM LEFT **The B250 was built in International's British factory. Here it is turning hay with a Lely rotary tedder.** *Peter Phillips*

ABOVE **The B250 was updated to the B275, then later to the B414 as pictured here.** *Ben Phillips*

RIGHT **The IH 454 signalled a new era for the company's tractors.** *Ben Phillips*

practical and was popular in Great Britain, as were the McCormick tractors, which were also produced under the International Harvester umbrella (and are described in their own sections).

In Great Britain International set up tractor factories at Wheatley Road, Doncaster, and in Bradford. The latter opened in 1954, some five years after Wheatley Road, and was the smaller of the two. It was where Jowett cars and vans had once been made, and here International produced the B series. The B250 tractor, introduced in 1956, was fitted with a four-cylinder diesel producing 30hp. It was a medium-range machine and was in an extremely competitive class of tractor, used by most farmers in the 1950s. The B250 was quite popular but was always a little behind the competition – having said that, in the five years it was in production around 30,000 were built. In 1958 the new B275 was seen as the B250's big brother, and had an increase of power to 38hp and slight styling changes. These tractors had small McCormick badges above the bigger International ones, but they have always been known as Internationals.

The early 1960s were good for the International brand, with the first Doncaster factory in Wheatley Road being joined by another in Carr Hill. Slight changes were again made in the new B414: the expanse of red paintwork of the B250 and B275 was broken up, the wheels, grille and side mesh now being cream in colour. The front grille was also different from the B275, which was a copy of that of the Raymond Loewy-designed Farmall tractors; the slats were now vertical instead of horizontal. During the late 1960s and 1970s International tractors built in Great

Britain became decidedly squared off, and the McCormick name was slowly being made smaller and smaller; when the B624/634 was launched it was the last tractor to be produced with the McCormick name.

International Harvester's main rivals had for years been building 'world' tractors, and in the 1970s the 454, 474, 475, 574 and 674 models were truly classed as tractor models for the entire world. This was a time when more sophisticated transmission systems were being developed, and International soon adopted both synchromesh and hydrostatic gearboxes – it was the first company in Europe to offer the latter.

The 84 series was built in Doncaster, entered the market in 1977/78 and became very popular. As with a lot of tractors around that time, a flat cab floor was offered. Some of the early International tractors based on the B250/275 had skid units and were fitted with a cab that tended to be extremely awkward for the driver's feet, so having a totally flat floor was a welcome feature.

Unfortunately, International Harvester needed to save money – the models, although popular, were not making great profits for the company. After a raft of cost-cutting measures were implemented, the workforce rebelled and a long strike was called, which left the company almost on its knees financially. There was only one way out, and that was to sell the agricultural side of the business. After much negotiation, in November 1984 Tenneco, an American gas and oil company, purchased International Harvester; it already owned two other major tractor manufacturing companies, Case and David Brown. From now on the company would be known as Case IH and was the world's second biggest agricultural equipment manufacturer. The Doncaster factory continued to produce Case IH tractors for the next 15 years, and to this day the famous 'IH' logo still appears next to the Case name, even though the company has changed ownership again.

ABOVE **The 84 series Hydro was a great success for some farmers.** *International*

BELOW **By the time the IH 584 was introduced the company was in need of help. This example is seen here baling hay with a New Holland 570 conventional baler.** *Ben Phillips*

JCB

JCB is somewhat of a British engineering success story. Joseph Cyril Bamford used his initials when he started his business in 1945, renting a small garage in Staffordshire to build trailers from post-war scrap. So popular were these trailers that by the end of 1940s he was employing six people. A that time he fitted a hydraulic unit to a trailer to create the first hydraulic tipping trailer in Europe.

In 1953, and using the company's experience with hydraulic tipping trailers, what was commonly known as a backhoe loader was fitted to a Fordson Major as a base unit, and the famous JCB badge was fitted for the first time. This model soon paved the way for the hydra-digger, which was produced in the now famous JCB colour of yellow. Although these early models were based on a Fordson Major, which was a popular farm tractor in the 1950s, it took until 1991 for JCB to produce what could be classed as a farm tractor. By now the company had made millions of items of plant equipment, which was exported to all four corners of the world – even if you weren't into diggers and excavators, you knew what a JCB was!

When it was introduced onto the market in 1991, the Fastrac filled a gap that existed for a fast versatile machine that could also act as an agricultural tractor. The farming industry embraced this concept and it became an instant hit. Its popularity was due to a

ABOVE LEFT **JCB began by building implements for existing tractors, such as this front loader fitted here on a Fordson E27N tractor. They would go on to use more Fordson models in the future.** *JCB*

ABOVE RIGHT **Other implements were also made, such as this mid-mounted mower fitted to a Fordson Major.** *JCB*

BELOW **When JCB introduced the Fastrac in 1991 it filled a gap in the market.** *JCB*

ABOVE **This new Fastrac tractor was certainly packed with a lot of technology, as seen in this cutaway diagram.** *JCB*

LEFT **Regular updates to the Fastrac range over the years have kept it looking fresh and modern.** *JCB*

number of factors; these included front and rear suspension, which provided unbelievable levels of traction and comfort for the centrally positioned driver. The central cab also gave a 50-50 weight distribution and was stable on most terrains, including the road. The fast road speed was also a big selling point, as the tractor could haul a trailer relatively easily, replacing the need for the lorries that were sometimes needed to transport equipment and produce as most tractors of that time were not sufficiently fast or comfortable. Of course, the Fastrac's style had been seen before with the Mercedes MB-Trac, which had been discontinued around the same time JCB introduced its new machine. The instant success of the JCB model showed that Mercedes-Benz had been wrong to drop the MB-Trac, and that JCB had been right to develop the Fastrac.

Since its introduction the JCB Fastrac has grown in popularity, size and model range, but its core values have remained. The latest tractors are available from 160 to 306hp, and the added list of specifications includes advanced xenon lighting, touch-screen controls and Xtra drive, which together keep the JCB Fastrac a British tractor success story. The company has come a long way from trailers built from surplus scrap.

BELOW **The latest Fastrac has a long history that has seen it grow bigger and better over the years. The 4220 shown here is pulling a Grégoire Besson cultivator.** *JCB*

John Deere

It is hard to believe, looking at John Deere today, that it all started back in 1837 when 33-year-old John Deere open his blacksmith's shop in Illinois, USA. With skills in blacksmithing in great demand, Deere's hard work soon became well known and he had a great understanding of the problems farmers were having with ploughs on heavy soil. He soon saw that a plough that was shiny and had a different shape from the traditional examples being used should work better in heavy soil, and he was right. By 1841 100 ploughs were being bought every year, and this increase in demand saw Deere enter an agreement with Leonard Andrus, who was able to provide much-needed help in producing larger quantities of these new ploughs. Five years later John Deere moved to Moline and left behind any link with Andrus; the move was made not only because of the geographical advantages of Moline, but also because of the more modern ways of making such equipment. Moline was to prove a good location for Deere throughout the remainder of his life, until he gradually gave his son Charles the reins of the business, which was now diversifying into many other tools associated with his plough. After an active life that saw him become mayor and play active roles in both the church and the bank in Moline, John Deere died in 1886.

At the onset of the 20th century the company was making quite a range of implements and soon saw that it needed to start producing tractors, so when Charles Deere died in 1907 his son-in-law, William Butterworth, took over the helm. He produced a few tractors, none of which were particularly successful, so when in 1918 an opportunity to purchase the Waterloo Gasoline Engine Company of Waterloo, Iowa, arose, he saw this as a great way to produce tractors. This company already produced the Waterloo Boy, which in Great Britain was marketed as the Overtime tractor, and was essentially the first John Deere tractor. The Overtime came to Britain courtesy of the Overtime Tractor Company and was largely the same as the Waterloo Boy, but with a different paint scheme and decals. It cost £295 5s 0d, which gave you a 24hp machine running on paraffin.

Soon the John Deere name adorned tractors in the familiar green and yellow colour scheme, and machines such as the Model A

BELOW **The Waterloo Boy and Overtime tractors were almost identical, the Overtime being the British version. Here it is dragging a trailed plough.** *John Deere*

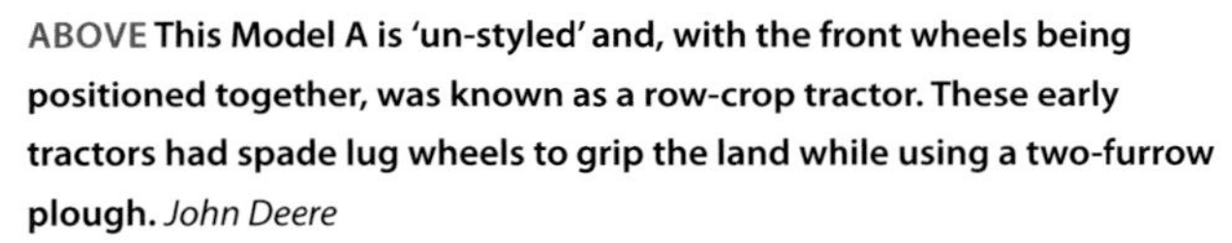

ABOVE This Model A is 'un-styled' and, with the front wheels being positioned together, was known as a row-crop tractor. These early tractors had spade lug wheels to grip the land while using a two-furrow plough. *John Deere*

ABOVE RIGHT The Model B shown here was 'styled' and featured nicely designed tinwork, as well as row-crop wheels. Although this tractor came to Great Britain this picture was probably taken in America. *John Deere*

RIGHT It was not really until the mid-1960s that John Deere entered Britain to any great extent, with models like this 4020 diesel. *John Deere*

and Model B could be found working in British fields thanks to the Lend-Lease scheme set up during the Second World War, which gave John Deere its first true path into Great Britain. These tractors were classed as 'un-styled' or 'styled'; if they were un-styled the radiator, fuel tank and steering mechanism were on show, but it was styled this same model was covered in by nicely designed tinwork to hide the aforementioned workings. A popular feature of many John Deere tractors of this era was the row-crop version, which saw the front wheels positioned together, fitted nicely between the rows of crops.

With John Deere being an American company, its presence in Europe was not as a good as it might be, so to gain greater access the company bought the German tractor manufacturer Lanz in 1956. These German tractors were then produced with John Deere Lanz badges, and had 'JDL' cast on the skid units, but the styling was very similar to the American John Deeres. These early models were not widely imported into the UK, but many have found their way into collections today. In the early 1960s another European factory, this time in Saran, France, started to produce the 303 and 505 models, which were the first European John Deere tractors without the Lanz name. The first two dealerships to start importing John Deeres into Great Britain were Norfolk-based Ben Burgess and L. E. Tuckwell of Suffolk, both of which continue to be dealers for the company today.

The 1970s and 1980s were a boom time for John Deere tractors in the British market, and in 1972 the company was awarded the Royal Warrant by Her Majesty the Queen. John Deere soon developed its distinctive cab, which featured a rounded door located at the front; this was unique amongst the squared-off offerings of its rivals. The cab was designed for safety, but also with comfort in mind, which was in contrast to many of John Deere's rivals, which just shoved a metal and glass frame on top with little regard for the driver's conditions. With cab innovations called 'sound-gard' selling well, John Deere tractors were gaining a reputation for being strong, well-built and well-equipped.

ABOVE The 3130 went through a couple of design incarnations during its production years from 1973 to 1979 – this is an early example. *John Deere*

RIGHT The John Deere 2130, produced 1973-77, was less reliable than many of its competitors with electrical problems and oil leaks. Today, however, this is a greatly sought after classic. *John Deere*

By the 1990s tractor manufacturers had been through hard economic times, which had seen many of them either merge or disappeared altogether, and while John Deere had not been immune to this tough financial period, it had taken steps to survive and came out stronger. This allowed the company to develop a whole new tractor range, the 6000 series, which was introduced into Great Britain in 1992. It soon garnered a number of awards for its innovative design and set up the company for the next decade. In 1998 it was announced that John Deere held the highest market share in Britain; this was the first time it had been in this position and came as a big boost to the company.

With the new century came an emphasis on satellite guidance fitted to tractors, so John Deere fitted GreenStar systems with Starfire to position the GPS signals. Automatic steering was also introduced, known as AutoTrac. Production for the tractors destined for Great Britain took place at the German factory in Mannheim, although many models did make their way over from Waterloo in America.

These days John Deere tractors are some of the most popular available in the UK and have a reputation for excellent quality; there are few farms that have not had a John Deere working on them at some point. The latest model range, the 9000, which includes tractors fitted with rubber tracks, are some of the biggest available, and have come a long way since that 19th-century blacksmith's shop!

ABOVE **The shape of the John Deere 1030 from 1979 became very familiar over the next decade. As with most John Deeres, this model was better equipped than its close rivals. Here a 1030 turns hay with a Fella hay rake.** *Ben Phillips*

BELOW **This John Deere 4240S produced 130hp from its six-cylinder diesel engine, and featured one of the best cabs available. The implement being used here is cultivating the soil to break it down to a fine tilth.** *John Deere*

RIGHT The John Deere 50 series was introduced in the mid-to-late 1980s. This tractor was capable of demanding work such as ploughing this heavy ground. *John Deere*

BELOW A cutaway picture of a John Deere 3350, made in Germany and fitted with a 6-litre six-cylinder diesel engine. *John Deere*

ABOVE The new John Deere 6000 series became very popular in Great Britain, helping it to become the best seller in 1998. *John Deere*

LEFT As with the 6000, the 7000 series featured the new 'comfort-gard' cab, which was not only quieter than the previous cabs but also offered greater visibility. *John Deere*

ABOVE The latest John Deere 6210R carries on the success of previous models, fitted with a 6-litre 240hp diesel engine. The implement being used is a set of discs to loosen the soil. *John Deere*

LEFT With a massive 9-litre six-cylinder diesel engine, the John Deere 8530 was a large tractor, capable of pulling this massive Vaderstad drill. The dual wheels on the tractor lessen the chance of the soil being compacted. *John Deere*

TOP RIGHT Introduced in 2015, this John Deere 9620RX has a 15-litre Cummins diesel engine producing 620hp and a Command View 111 cab. Just one from a whole new range of big John Deere tractors. *John Deere*

BOTTOM RIGHT This John Deere 7310R features the latest technology including touch-screen controls and hydraulic joystick, LED lighting and the Command View 111 cab. *John Deere*

9620 RX
JOHN DEERE

Kubota

Kubota was formed in Osaka, Japan, in 1890 and originally made cast-iron water pipes, after which it turned to making motorised products. It has been a major force in Great Britain since it set up a base in Thame, Oxfordshire, in 1979. This central location was specially chosen to make it easily accessible to everywhere in the country. The company became famous for producing bright orange compact tractors that soon gained a reputation for being strong and reliable machines.

Although the company's roots lie firmly in the Far East, and its early models have very much a 'Made in Japan' look, these tractors have a lot to thank Great Britain for. Since Kubota set up shop here it has developed many 'firsts' in the agricultural industry; a compact tractor fitted

LEFT **Kubota built a strong reputation on little tractors such as this B6000. It featured four-wheel drive and from this tractor a new market was created by the Japanese company.** *Kubota*

BELOW **This early Kubota B6000 compact tractor, fitted with a front loader, made this little machine even more useful.** *Kubota*

LEFT What appealed to market gardeners, who represented Kubota's key market, was that the small and compact Kubota could work in a greenhouse. Here it is using a rotavator to work the soil. *Kubota*

BELOW Kubota has not forgotten its rich heritage in compact tractors – these little orange machines can be found all over Great Britain. They still use the B prefix and shown here are the B3150, BX2350 and B1620. *Kubota*

with a four-wheel-drive system was one such innovation, together with hydrostatic transmissions and power steering, to name just a few. Mastering these systems on a compact scale allowed Kubota tractors to become extremely popular; as other major manufacturers were making ever bigger machines, they forgot that there was a place for smaller, rugged tractors. Kubota capitalised on this and filled the gap in the market; it could be said that the company brought compact tractors to the masses.

Over the years Kubota has not only carried on making great compact tractors ranging from 16 to 31hp, but has also increased its line-up to include models right up to 175hp, which now makes for a comprehensive tractor range. As well as tractors, the company produces ride-on mowers, utility vehicles and construction equipment. Kubota can easily boast that it produces the best-selling compact tractor in Great Britain today; the high standards in manufacturing for which the Japanese are renowned are clearly seen in these popular orange machines. Further proof, if needed, is the strong resale values for these tractors.

BELOW **Kubota has kept its designs fresh over the years and this M9960 shows how good they look.** *Kubota*

TOP RIGHT **Today Kubota has increased its range to include bigger and more powerful tractors. This includes the M7151 (left), which has a 150hp diesel engine, and the M7171, which has a 170hp 6.1-litre diesel.** *Kubota*

BOTTOM RIGHT **Having the M7171 has put Kubota into new territory in the tractor market and not one you expect from this company which was so strong in the compact area for many years.** *Kubota*

KUBOTA
KUBOTA

KUBOTA

Leyland

British Leyland will probably be remembered for all the wrong reasons, mainly strikes and building cars that were not of the best quality. It was a massive company in the 1970s, producing everything from Jaguar to Rover cars in the Midlands. However, further north in Bathgate, Scotland, it was also building two-tone-blue tractors. In 1968, when the British Motor Corporation (BMC) merged with Leyland Motors, the British Leyland badge appeared on the production line. BMC had been producing Nuffield tractors in poppy red for years, so now with a new name came a new colour scheme and some new tractors. As with most tractor manufacturers of the time, Leyland owned a 150-acre farm next to the factory in which to base a training centre. It was also the backdrop to many publicity photos the company used in its sales literature.

The smallest of these tractors was the Leyland 154. This had been in production for years during BMC days, and with a new name there came a few minor changes to this little machine. It was fitted with a 1.5-litre four-cylinder diesel, hence the 154 name. This mid-20hp tractor was more of a smallholder's machine, or could be fitted with a set of gang mowers for a golf course or playing field. In 1971 Leyland went into straight competition with Massey Ferguson and its hugely successful 135 model by launching the 253. Even though it had the same Perkins engine as the MF, it was never going to trouble the dominance of the 135.

A year later the engineers at Bathgate introduced a whole new engine for Leyland tractors, and with it a new model numbering system. The new engines were designed to hopefully address the problems associated with the old ones, which had been carried over to certain models from the BMC era. Over the years this new Leyland engine was deemed a great success. The need for safety cabs was becoming ever more essential due to the British Government clamping down on accidents involving tractors rolling over. Some companies, such as Massey Ferguson and John Deere, which were big rivals of Leyland, had embraced these rules, and until recently Leyland had fitted cabs that were just metal frames. In 1976, to comply with yet more laws governing sound inside cabs, Leyland introduced the Q cab.

A synchromesh gearbox had been in development for many years and was finally fitted in 1978. The models so fitted had 'Synchro' in bold red letters across the silver decal applied down the side of the bonnet. This was a real step forward in engineering, and what made it even better was the fact that the gear levers could be positioned to the side of the gearbox instead of on top, allowing the cab floor to be clutter-free. Around this time the Italian firm

BELOW **In 1970 the Leyland 154 was the smallest tractor in the range and was carried over from the previous model BMC Mini 9/16.** *Ben Phillips*

TOP RIGHT **The Leyland 2100 was introduced in 1973, fitted with a six-cylinder engine.** *Nuffield and Leyland Tractor Club*

CENTRE RIGHT **The Leyland 4100 tractor had four equal wheels. Not many were built, making them quite rare today.** *Nuffield and Leyland Tractor Club*

BOTTOM RIGHT **1976 saw the introduction of the mid-range Leyland 272 with a four-cylinder engine.** *Nuffield and Leyland Tractor Club*

Carraro supplied a four-wheel-drive system for Leyland tractors; earlier in the decade Leyland had sourced County front axles for a four-wheel-drive solution.

As the 1980s began Leyland wondered if a change in direction regarding colour scheme and numbering system should be adopted. Its existing models were looking tired and dated compared to those of many other tractor manufacturers. After experimenting with a number of colour combinations, 'golden harvest' was applied to the bonnet and wings, and the skid unit was black; this was decided upon largely because no other tractor had this colour scheme and Leyland wanted a unique colour. The new numbering system included the models 502, 602/604, 702/704 and 802/804, and together with the new colours and numbers an updated specifications list was also added. Most of the new or updated features were not fitted on the 502 model, but the rest of the range had several innovations. These included a new cab built by Danish firm Sekura, which delivered a rather smart environment for the driver to sit in, known as the Explorer cab. The normal Leyland QM cab was improved with lower noise levels inside as the Government continued to demand better cabs that were safer and quieter. The brakes were also updated and now had a multi-disc that was immersed in oil, and a stronger linkage was fitted on most models.

The Leyland 154 had been dropped many years previously and replaced by the 235; this was made in Turkey and never officially appeared in Britain; however, shortly after the updated range was introduced it made a reappearance as the Leyland 302.

In 1982 things were looking good for Leyland tractors: the updated range looked good and most observers considered this to be one of the best periods the company had seen. So many were amazed when Leyland sold its entire tractor concern to Marshall; maybe the company knew that the good times would not last and got out while the

LEYLAND SYNCHRO 482

LEYLAND
LEYLAND

TOP LEFT The Leyland 482 Synchro was identified by the red writing on the grey decal strip. *Peter Squires*

BOTTOM LEFT Leyland demonstrated the Synchro models in a muddy field to show off their capabilities to the best effect. *Nuffield and Leyland Tractor Club*

ABOVE Leyland tractors were sold to Marshall and both are still found today, kept in service by enthusiasts of the brand. *Peter Squires*

BELOW In the 1980s a new colour scheme was introduced called 'golden harvest'. *Nuffield and Leyland Tractor Club*

going was good. From mid-January 1982 no further tractors were produced in the Bathgate factory, and by the end of the month the sale had gone through.

The whole business of British Leyland workers striking had affected both car and tractor production, and some thought this was why the tractor side was sold off quickly, perhaps too quickly and cheaply. The loss of tractor production did nothing for the prospects of the Bathgate area, although truck production went on until 1985.

Marshall

Gainsborough in Lincolnshire was a location that became famous for producing Marshall tractors. Before that, however, the factory had been synonymous with the production of steam engines. The site had been acquired in the early 1840s by William Marshall, and he began to produce stream engines in 1849. By 1861 William's sons James and Henry had both become partners, and by the turn of the century tractors were firmly on their mind. Seeing a single-cylinder tractor from Germany in the late 1920s gave them the idea of producing a similar machine; this was the Model E, then eight years later the Model M arrived.

Just before the end of the Second World War in 1944 a new tractor appeared, known as the Field Marshall. The design was familiar to the previous tractors but featured more modern styling, which was rounder and smoother; however, under the newly styled body the mechanics were very similar to the previous models from the 1930s. The Field Marshall went through four revisions, known as series 1, 2, 3 and 3a, each update improving a little on the original. However, it took until the series 3 to finally iron out the problems that sometimes dogged the transmissions. The later models were painted orange, which replaced the familiar green found on the early examples, but by the early 1950s the Field Marshall was unable to keep pace with its rivals.

LEFT **The Marshall Model M was introduced in the late 1930s, after the E.** *Ben Phillips*

BELOW LEFT **The Field Marshall was probably the most famous tractor the company produced, and went through four updates.** *Ben Phillips*

BELOW **Following the previous green Marshall tractors came this bright orange MP6 complete with Leyland six-cylinder engine.** *Peter Phillips*

TOP RIGHT **Many Marshall MP6 tractors went to Australia, but on a whole this model was considered a flop.** *Peter Phillips*

BOTTOM RIGHT **Looking for a way back into tractor production, Marshall bought Leyland Tractors and put Marshall decals on the bonnet. This Marshall 802 is pulling a grain trailer.** *Peter Squires*

MARSHALL
MARSHALL
802

804
MARSHALL
MARSHALL 804

954
MARSHALL
MARSHALL 954 XL

TOP LEFT **The old Leyland designs were becoming tired and, with lack of money to develop new models, the writing was on the wall. This pair of 804 four-wheel-drive tractors is pictured in the 1980s.** *Marshall Tractors*

BOTTOM LEFT **Rebadging Steyr tractors gave Marshall a more modern product to sell, as shown here on this 954 model using a three-furrow plough.** *Peter Squires*

ABOVE **Unfortunately the rebadging was not enough and Marshall tractors disappeared, together with the factory. A Marshall 115 is seen here pulling a trailer containing sugar beet.** *Marshall Tractors*

There's nothing quite like the noise of a Field Marshall thumping away – the 'bomp bomp' sound is mesmerising, as is the way the engine makes the tractor bounce slightly on the spot. It is not hard to see why Field Marshalls are so highly regarded by collectors and enthusiasts, a fact highlighted by their relatively high price-tag.

After the Field Marshall the company produced a model called the MP6. This bigger machine was finished off with bright orange paintwork like the final Field Marshalls. Power came from a Leyland six-cylinder engine, and not many were built; in fact, the total was only 197. Many of them were exported to Australia, very few were sold in Great Britain, and the model was considered something of a flop. In recent times some Marshall MP6s have found their way back to these shores and have ironically become a very sought-after tractor amongst collectors; in 2015 an example fetched £80,500 at a British auction.

For many years the Marshall name largely disappeared from tractor production when it decided to develop a range of crawlers known as the Track Marshall. In the early 1980s, wanting to get back into manufacturing tractors, the company bought the agricultural section from British Leyland; this largely comprised Leyland tractors, which were painted in 'golden harvest' and had full Marshall badges and decals applied.

Unfortunately, due to lack of funding Marshall failed to develop new tractors and their appeal to farmers dwindled, so in 1991 production stopped. These ageing models simply could not compete with the major tractor manufacturers of the time, so trying to keep up Marshall started rebadging Steyr tractors. This did not work either, and within a few years the end came for Marshall tractors and the famous Gainsborough works closed; a shopping centre now adorns the site.

Massey Ferguson

When people think of tractors, most think of a bright red Massey Ferguson. The name Massey Ferguson came about when Massey Harris merged with Ferguson, creating one of the most famous brands in the agricultural world.

Massey Ferguson was founded in 1958, during the production of the 35 and 65 models. Strictly speaking, the merger had happened earlier in 1953, but was at first known as Massey Harris Ferguson. The 35 model had been produced under the Ferguson name and was grey and gold in colour; fitted with a four-cylinder diesel engine, it was not proving a great success. Its poor starting ability was giving the model a bad name, so at the end of 1959 it received a new engine and a new colour scheme. The engine was made by Perkins and proved not only better at starting but was also powerful, reliable and set to power these tractors for decades. The new red and grey paintwork was also set to last for years and become synonymous with Massey Ferguson.

The 1960s were to prove an important decade for the company; when it started only a few tractor models made up the range, two from the Ferguson side and the rest from the Massey Harris era. By 1964 the MF35x and 65 Mk 2 were probably the best and most popular tractors on the market, but it was time for a change, and what Massey Ferguson had planned really set up the company as a world-beater. It brought out the 100 range, nicknamed 'The Red Giants'; the 130, 135, 165 and 175 were the first models to be introduced, followed later by the 185. They were all painted with red tinwork and grey castings. The shape had a family look about it – all models were largely the same but with subtle differences – and they were all fitted with three- or four-cylinder Perkins diesel engines. Perkins was now owned by Massey Ferguson, so the engines were designed and built exactly to the parent company's specifications, having been developed over the years and widely acknowledged as the best. Of course, Ferguson was famous for the hydraulic system it had developed and brought to the tractor market in the 1940s, and considerable tweaking and updates kept Massey Ferguson ahead of the game. By the end of the decade it had grown into an expanded range of tractors, and as the 1970s started the 100 range was updated.

ABOVE **Massey Ferguson is probably the most recognisable tractor and other agricultural machinery brand in the world. In this picture taken in the early 1960s, probably not far from the factory in Coventry, there is a Massey Ferguson 35 baling and a Massey Ferguson 400 combine working away.** *AGCO Archives*

RIGHT **Massey Ferguson's products were soon recognised by their red and grey colours. This Massey Ferguson 35 has been restored to its condition when it came out of the factory.** *Ben Phillips*

During the 1970s Massey Ferguson was the biggest tractor manufacturer in Europe, America and Canada. It had factories all over Great Britain; the main one was the famous Banner Lane plant in Coventry, but there were others in Manchester, Liverpool and Kilmarnock, making everything from tractors to construction equipment and combine harvesters. There was also the Perkins engine plant in Peterborough and training facilities, and with all these locations combined the total workforce was near 20,000.

As Massey Ferguson grew it started to build bigger tractors to cater for farmers who had increased their field sizes. The model that was introduced was the 1200, manufactured in the Manchester factory. It was fairly revolutionary in that it featured an articulated chassis. Although Massey Harris had dabbled with this kind of tractor design with the General Purpose model in the 1930s, with limited appeal, 40 years later it was assumed that it might be more successful. When designing the 1200, Massey Ferguson certainly thought it through: the cab was very modern in design and featured a great layout, including controls that were easy for the driver to find quickly. The seat was comfy and the expanse of glass gave an excellent 360-degree view; it was also one of the first Massey models to allow for a radio to be fitted. The engine was a six-cylinder Perkins and gave the tractor a power output of around 100hp; this was somewhat of a milestone as many tractor manufacturers were striving for this kind of horsepower.

At the end of the 1970s Government legislation was passed that meant that tractors now needed properly enclosed cabs that were

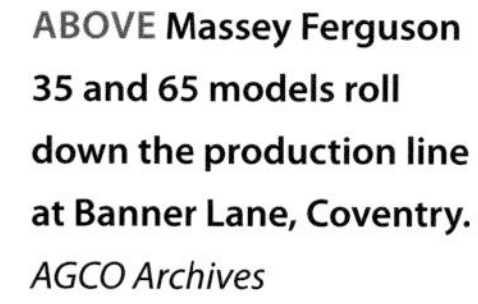

ABOVE **Massey Ferguson 35 and 65 models roll down the production line at Banner Lane, Coventry.** *AGCO Archives*

LEFT **The 35's bigger brother was the 65, which featured a long bonnet with a four-cylinder Perkins engine beneath it.** *AGCO Archives*

not only safe but also quiet – in fact, less than 90dB. To address the new laws Massey Ferguson brought out the 500 series, which was designed with a cab in mind.

The hydraulic system was further improved and the controls were moved inside the cab so they were easier to use. It was also easier to mount the tractor as it was fitted with a flat floor; no longer was there a gearbox to straddle, which was sometimes tricky when wearing boots plastered in mud.

The 200 series was the direct successor to the 100 series, with the small models in the range being numbered 230, 240 and 250. Under the new restyled tinwork they were basically very similar to the 135 and 148 models; even the Perkins AD3-152 diesel engine was very similar, if not the same. The bigger models featured the Perkins AD4-236 diesel engine, so the 265 was the updated version of the 165. Although these new models may have had very similar engines to those that they replaced, the bodywork was new and very similar to the 500 series, and this new design set the style for the 1980s.

A vastly improved cab was fitted to the 600 series, which Massey Ferguson intended to be seen as a classier tractor than the 500 series it replaced. John Deere had for a while been offering well-equipped, comfortable cabs, and Massey Ferguson had fallen behind in grasping this much-needed addition. Perkins engines again powered this range of tractors; the 675 had the tried and tested A4-236, which produced 66hp, while the range-topping 699 had the A6-3544 at 95hp. These models were built in Great Britain and France, but unfortunately did not sell as well as previous Massey Ferguson models; this was blamed on a poor design that failed to attract farmers, or maybe that the cab was too high and limited its use around low buildings on the farm.

In 1986 the 3000 series was introduced, a new model range that embraced the new technological age. It changed everything – never before had electronics controlled so much on a tractor. With 'Autotronic' printed on the door, this meant that the PTO, four-wheel-drive system and diff-lock had a degree of automatic control. On the other hand, if 'Datatronic' appeared on the door, drivers had a large amount of data presented to them on a digital display in the cab.

BELOW **The 65's styling was taken from the American-built Ferguson 40. Here this one is undergoing field tests in Coventry in the late 1950s using a Massey Harris baler.** *AGCO Archives*

RIGHT **The Massey Ferguson 100 range was launched to a big fanfare and dubbed 'The Red Giants'.** *AGCO Archives*

BELOW **The 135 was the most popular tractor in the 100 range. This early model, recognisable by the red wheel centres and the round shell fenders, is seen here using a Massey Ferguson baler.** *AGCO Archives*

165
CRW701C

165

TOP LEFT **At the bigger end of the 100 range was the 165, which effectively replaced the old 65 model.** *AGCO Archives*

BOTTOM LEFT **The 165 was updated throughout its life; most notably the Perkins four-cylinder engine was new. Here in this publicity photo, probably taken in Coventry, it is ploughing with a three-furrow Massey Ferguson plough.** *AGCO Archives*

ABOVE **In the late 1970s MF bought out the 500 series of tractors with enclosed cabs, with safety in mind.** *AGCO Archives*

If the 3000 series was bristling with electronics, the 300 series was a simple no-nonsense tractor that went on to be one of the most popular tractors ever made. It was fitted with either a Lo-Line or a Hi-Line cab, which made the range useful and appealed to many farmers and horticulturalists. The cab was bright and airy, and was finished in a light grey colour, which was nice when new but soon became grubby. The front axle featured two- and four-wheel-drive options, all with hydrostatic steering, which gave unparalleled control. The Perkins engines had power outputs from 52 to 100hp. The 3000 and 300 series were the first Massey Ferguson tractors to have the round headlamps, which had adorned the front since the model 35 days, replaced by square ones. However, years later the round ones would return.

Massey Ferguson saw many owners over the decades. The Argus Corporation was one, and when Conrad Black took a controlling stake he oversaw a period that delivered profits thanks to a number of cost-cutting initiatives. But even under Black the profitable times could not last, and the company slipped back into losing money due to hard global economic times, especially in the tractor market. This led to the Canadian Government bailing it out, and Massey Ferguson was now owned by the Varity Corporation.

Under this new management many more revenue-losing areas were identified and dealt with, largely by selling them off. This steadied the ship and Massey Ferguson started to trade more successfully again. Now that the company was selling more tractors than its closest rivals it was bought by the AGCO Corporation. This big American company, set up from a management buy-out of the now defunct Deutz-Allis concern, saw its fortunes improve immediately as it now owned the most recognisable agricultural company in the world. Under new ownership, the Massey Ferguson range was reorganised and many of the very big tractors were discontinued. AGCO had other manufacturers under its umbrella and did not want Massey Ferguson to compete with them, even if the company was the jewel in its crown.

Following the success of the 3000 range came the replacement 3600 to 6100 models, which had power from 86 to 111hp. From a distance, the 6100 looked very similar to the models it had replaced;

The 1200 gave the company a tractor in the 100hp bracket, and it has become an iconic, classic machine. This 1200 is pulling a set of Massey Ferguson discs in a field that has previously been ploughed. *AGCO Archives*

MF
1200
MF

it was also not greatly loved by many farmers, who reported problems that they had not come to expect from this brand. The company addressed the situation when it launched the 6200 range, which was a better tractor; many were relieved, as it had been a long time since Massey Ferguson had made such an unloved model. The 6100's bigger brother was the 8100, and the emphasis on electronics continued with Datatronic II, which gave more information through its 22 separate functions.

The small models included the 4200, which had the sloping bonnet that was becoming all the rage around this time. With the bigger tractors having more powerful engines, they needed longer bonnets, so the visibility of the front of the tractor was greatly reduced. The smaller tractors had a compact Perkins engine, so a lower front end could be developed and used around the farmyard with ease. The low-profile cab also added to the versatility of this model range, being particularly useful under low buildings or tree branches.

Mention Massey Ferguson tractors, and most farmers and enthusiasts will probably think of the Banner Lane factory in Coventry. This famous plant had been producing tractors since the first Ferguson T20 model had rolled down the production line in 1946. Over the years some of the most iconic tractors had been designed in one of the many offices, and the factory had turned them into reality, it had kept the area's employment buoyant, and other local companies benefited from the added work in supplying components. However, AGCO was looking to close either the Banner Lane factory or the one in Beauvais in France, as it was trying to streamline the company, and one factory in this part of Europe would suffice. Beauvais looked safe as the French Government put pressure on AGCO to keep it open; that worked, but signalled the end of Banner Lane, which closed in December 2002. This was a harsh blow to the area and most people connected with agriculture found it sad that one of the last big tractor manufacturing plants in Great Britain, with such a rich history, had closed. Production in France increased to make up for the Banner Lane closure, which was destined to become a housing estate.

LEFT The 565 model was a good all-round tractor. This 565 is pulling a three-furrow plough while undergoing field tests and publicity photos. *AGCO Archives*

ABOVE The 575 model was at the bigger end of the range, and is considered a classic today. This example is pulling a Massey Ferguson 128 baler. *AGCO Archives*

Today Massey Ferguson, still under AGCO ownership, continues to be as loved as ever; even through tough economic times Massey Ferguson tractors have remained popular. A model range that appeals to dairy and arable farmers will always sell well in Great Britain, and the 5400 range does just that. Great ground clearance, unparalleled views from the cab and a compact design tick all the boxes. Recently the most famous new model range has been the 5600, an example of which the company took to Antarctica. Back in the 1950s a group of Ferguson TE20 tractors trekked across the frozen continent, and the company wanted to go back, so in 2014 a fully-equipped 5610 set off and proved once again that an

agricultural tractor could cope with extreme weather conditions not normally found on the average British farm. The trip was dubbed 'Antarctica 2', and for 2,900 miles a Massey Ferguson 5610 drove in probably the harshest conditions found on earth. The tractor was equipped with special tyres to cope with travelling over snow at speed, and the cab was fitted with extra heating systems. The glass was also replaced with polycarbonate, which was less likely to break in the wind. The trip was as much of a success as the first one, but this

LEFT The 200 series was considered to be the 1980s equivalent of the 100 series from the 1970s, and under the modern tinwork this 240 certain looks very similar to the 135, but obviously with updates. *Ben Phillips*

BELOW The 600 series featured smarter cabs than those of the 500 series. This example being four-wheel drive, it can work on heavy land such as here with a set of discs. *AGCO Archives*

The MF 2625 had electronics fitted, setting the scene for the models that followed. This one is probably working on a farm local to the factory in Coventry. *AGCO Archives*

time, due to greater media presence, it was better known. Hot on the heels of this success, Massey Ferguson bought out a 5610 Antarctica 2 special edition, which had added features such as black cooling grills, special decals and each one individually numbered.

Massey Ferguson continues to be the most recognisable tractor brand, and rightly so. Over the years it has had many owners, but under AGCO it seems to have an extremely secure future. Harry Ferguson would surely approve of most of what Massey Ferguson represents today; however, he probably wouldn't like the fact that they were no longer made in Great Britain.

LEFT **The 3000 range changed everything – many of its functions were now reliant on electronics.** *AGCO Archives*

BELOW **Two- and four- wheel-drive 3000 models were available. This 3050 is using a spring tine harrow to weed around a crop of young maize.** *AGCO Archives*

OPPOSITE **Most of the 3000 tractors sold were four-wheel-drive. This 3090 datatronic is using a rotavator to break the soil down prior to planting the crop.** *AGCO Archives*

3090
MASSEY - FERGUSON
datatronic

ABOVE The 300 range offered either the Hi-Line or Lo-Line cab. This is a Hi-Line example. *AGCO Archives*

RIGHT The 300 range was a lot smaller and did not have the fancy electronics of the 3000. This small 390 is using a set of gang mowers on a golf course or public park. Note the special grass tyres fitted to avoid damaging the surface. *AGCO Archives*

MASSEY-FERGUSON 360

LEFT Square headlamps replaced the more usual round ones, a sign of the times in which this tractor was made. This 390 has a rotary hay turner fitted and is working in a field at Trimpley in Worcestershire. *Ben Phillips*

BELOW This MF 5610 made it all the way to the South Pole in 2014, just as a group of Ferguson T20s did in the 1950s. This expedition was called 'Antarctica 2'. *AGCO*

RIGHT **One of the latest mid-range tractors built by Massey Ferguson is the 6616. This is a popular choice with farmers in Great Britain, including this example in Trimpley, Worcestershire. Here it is mowing hay with a Kverneland mower.** *Ben Phillips*

BELOW **The MF 8700 range is the biggest on offer today, with models right up to 400hp.** *AGCO*

Massey Harris

Thinking of Massey Harris as a foreign firm because it originated in Canada is wrong, as in those days Canada was part of the British Empire. If only Daniel Massey had known in 1847, when he had a modest workshop in Ontario, that one day his forename would become part of the most recognisable name in agriculture! In his workshop he produced farm equipment that was simple and easy to use, but he had fierce competition in the same area of Canada from Alanson Harris. As they were both in the same business, instead of trying to compete with each other they merged in 1891 and the company was known as Massey Harris. It was now the biggest machinery producer in Canada, a country that had a large amount of farmland needing equipment that Massey Harris was well placed to supply.

Surprisingly it took until 1917 for Massey Harris to enter the tractor market in Great Britain, with its first model known

ABOVE The 1930 General Purpose tractor was one of the first true Massey Harris machines to be produced. *AGCO Archives*

RIGHT In 1939 the Massey Harris 101 Junior was introduced, and featured bright red bodywork and yellow wheels. *Ben Phillips*

as the Whiting Bull. This three-wheeled machine with a 20hp twin-cylinder engine proved capable of ploughing a 10-acre field in Kent with relative ease. The success of the Whiting Bull was not to last long, however, largely due to a shortage of parts, and a year later production ended. Massey Harris was now without a tractor, but the interest gained from the Kent ploughing demonstration with the Whiting Bull convinced the company to enter an agreement with a tractor company called Parrett. This was not to last long either, as the Parrett tractor was somewhat dated and, with key rivals producing more modern machines, not many were sold. Finally, after two failed attempts to enter the tractor market, joining the Wallis Tractor Co from Wisconsin was to provide the desired success. The first fruit of this new venture was the Wallis Cub, and things went so well that by the end of the 1920s Massey Harris had taken over the Wallis company and rebadged the tractors as Massey Harris.

The first model built without another tractor company's help came in 1930, and was known as the General Purpose tractor. This

ABOVE **The 744PD was the first Massey Harris tractor to be built in Great Britain, and featured a Perkins diesel engine. This painting by Terence Cuneo shows this tractor perfectly.** *AGCO Archives*

was probably too revolutionary for its time, having a high clearance, four-wheel drive and an articulated chassis. Although it featured a lot of good ideas and was unbeatable on rough ground, sales were poor and production was cut short.

The participation of Massey Harris tractors in the British market really took off in the late 1940s when the company opened its first factory in Manchester. The first model produced was the Massey Harris 744PD, which was fitted with a six-cylinder Perkins diesel, hence the 'PD' in the name. The number 7 indicated that it was British-built, as the Canadian equivalent was just known as the Massey Harris 44. By now the colour scheme that Massey Harris had adopted was red and yellow, these vibrant colours replacing the drabber dark green for which the company's earlier tractors had been known. During the 744 production run, a factory in Scotland

leaders, namely Ferguson and Ford, which both had the revolutionary three-point linkage and fresher, more modern designs that the Massey lacked. So when Harry Ferguson approached Massey Harris offering to sell it the US side of his business, the company responded by proposing to merge both companies into one. In 1953 the deal was done, but instead of immediately renaming everything Massey Ferguson the company still produced tractors branded Massey Harris and Ferguson separately. If you looked closely at the Fergusons of that time, the letters 'MHF' – Massey Harris-Ferguson – could be found, so slowly the machines were becoming known by this future name. By 1958 the Harris part had been dropped, and now all products would be known simply as Massey Ferguson. Banner Lane in Coventry would become the company headquarters, and manufacturing facilities in France and Canada were just a few of those located around the world.

ABOVE **The smallest Massey Harris tractor was the Pony. Although not built in Britain, many have come into the country over the years. It was introduced in 1947 and production ran for about 10 years.** *AGCO Archives*

at Kilmarnock was opened, and the 745 followed shortly afterwards; again, as with the 744, a Perkins diesel was fitted, but this time the L4, which had four cylinders. Just over 11,000 of these tractors were built in Kilmarnock.

The smallest tractor Massey Harris produced was probably their most popular – the Pony. This little tractor was not made in Britain but in the French and Canadian factories. Many of Massey Harris's rivals were building small tractors, such as Allis-Chalmers and Farmall, and they were proving very popular. Until now Massey Harris had been building big, heavy tractors, so this new small lightweight machine was a new direction. The Pony sold best in Europe, especially France, as it was useful in the many vineyards for which rural France was famous. In recent years many Pony tractors have made their way into Great Britain, and imports from France have given many tractor collectors on this side of the Channel an opportunity to own this lovely little tractor that is full of charm. These French-built models are predominantly fitted with four-cylinder petrol engines made by Simca or Peugeot, but there was also a Hanomag diesel engine available.

By the early 1950s Massey Harris was struggling to keep up with the market

BELOW **The Pony was offered with a couple of petrol engine options as well as a diesel.** *Ben Phillips*

McCormick

ABOVE **Although this tractor is badged 'International W4', many were seen as McCormick W4.** *Peter Phillips*

The original McCormick tractors were produced as part of the International Harvester operation; they were branded McCormick Deering and marketed alongside International and Farmall. One of the founding members of International Harvester was Cyrus McCormick, so it is not surprising that his surname would find its way onto a group of tractors. One of the most popular McCormick tractors in Great Britain was the W6, introduced in 1940. This was a tractor much used for belt work; during this time a number of important agricultural implements relied on belts to power them, so tractors were fitted with a pulley to do this. The McCormick W6 can also be found bearing International decals, and nearly 30,000 were built during its 14-year production run. The W9, the W6's bigger brother, was actually more popular overall, with just over 67,000 units built. This was probably due to the American market, where it was made, which needed a bigger tractor to cover the large prairie fields. Great Britain had fields that were a lot smaller, so the W6 was more suitable. The Doncaster factory that made Internationals and Farmalls also produced certain McCormick tractors such as the Super BW-6, Super BWD-6 and the B450. Over the following years many International tractors had 'McCormick' in small writing on them, but the last model to have the McCormick name was the B-634. Production of this model ceased in 1972, and from then on all tractors were called International until finally renamed as Case IH.

RIGHT **The International W4/McCormick W4 can be found with either decals on the bonnet.** *Ben Phillips*

Little did anyone think that the McCormick name would ever grace a tractor bonnet again, but in 2001 it did. Case had always held the brand name McCormick but never used it, so when it merged with New Holland to create one of the biggest tractor manufacturers in the world, European law stipulated that Case get rid of some of its assets; it was feared that the merger would make the new company too big and powerful. Case agreed, and disposed of the Doncaster factory, certain tractors from the Case range, technical knowledge connected to these models, and of course the McCormick name. This then allowed Case to become part of New Holland, which was owned by Fiat. After a lot of negotiations with a number of different companies, ARGO S.p.a, an Italian firm that owned Landini amongst others, finally purchased the Doncaster plant. This famous factory became the headquarters of McCormick tractors and had an agreement to use Case Maxxum designs but with a Perkins engine fitted. Having the McCormick name back at Doncaster was nostalgic for many, even if it had only tenuous links to the original McCormick tractors.

It didn't take long for ARGO to reinstate McCormick tractors as an established brand; the name obviously helped, as well as the popular designs based on the Case IH tractors. In the late 1990s Case IH had built some beautiful tractors that farmers seemed to like, and McCormick was allowed to copy these for a couple of years, then it would have to design its own. The tractors are a brighter and slightly lighter red than the Case IH colours, and the wheels and grille are silver, which harks back to the original McCormick days. By 2007, with a complete range of newly designed tractors from 23 to 280hp and an impressive dealer network in 55 countries, the company invested £7.5 million in upgrading the factories in Italy. However, this was not good news for Great Britain, as production moved from Doncaster to these new facilities and the famous International, then Case IH, factory produced its last tractor. More than 60 years since the first tractor rolled down the production line, the last, a McCormick XTX215, came through in December 2007. The last McCormick CX model built there had a unique livery and was auctioned off for just over £31,000 to a Cambridge dealer who wanted to put it in his collection, which, judging by other special tractors, should be a sound investment.

LEFT **The bigger W9 was useful as a belt tractor, as the pulley on the side could drive many implements.** *Ben Phillips*

ABOVE The McCormick-badged 434 was largely an International B series under the tinwork. *Peter Phillips*

LEFT The McCormick 634 was the last McCormick-badged tractor made by International. *Peter Phillips*

Today McCormick has fresh new designs that are modern-looking and do not rely on old Case IH models. The range has now expanded and a number of vineyard tractors are produced in a new purpose-built factory in Italy, set up when the Doncaster plant closed. There is a McCormick for virtually any job, and even though the brand name was only reintroduced in 2001 and has no real direct links with the original McCormick company, it certainly has come a long way in a short space of time and is definitely worthy of its illustrious name.

Minneapolis Moline

The Minneapolis Moline tractor company was formed when three companies came together, based in Minnesota. The tractors were popular throughout America, and were made available to farmers in Britain thanks to the Second World War, when tractors were in great demand. Fordson did its best to supply as many Standard N tractors as possible, but to help with the shortage many machines were imported from America, Minneapolis being one supplier. Farmers had to apply to the Department of Agriculture and, if successful, generally had no say in what they got. Those who got a Minneapolis Moline were extremely happy as the tractor came with rubber tyres, which, strange as it seems today, was a luxury. The Model U, the principal model imported, was a lot faster than the Fordson N, and was fitted with a 4.6-litre four-cylinder gasoline engine producing 27hp.

Although Fordson had changed the colour of the N to a dark green in the hope that enemy planes would not spot it working in the fields, the Minneapolis tractors were bright yellow, a colour known as 'prairie gold'. Farmers who took delivery or this imported tractor saw the bright colour as a ray of hope in the dark days of war, and as such they treasured the machines more than they probably would any other tractor.

After the war very few Minneapolis Moline tractors came into Great Britain, but those that had come over during the war had earned their place in British tractor history. However, the company was in trouble; its factory workers were embroiled in all sorts of disputes, and strikes were inevitable. In 1963 Minneapolis Moline was bought by White, another American tractor producer, and a

TOP RIGHT **The Minneapolis Moline UTS came to Great Britain during the Second World War.** *Ross Bartlett*

BOTTOM RIGHT **The big MM Model G was made between 1938 and 1959.** *Peter Phillips*

BELOW **This Model R was the row-crop version, and one of the most popular MM tractors.** *Ben Phillips*

little over 10 years later the name was dropped. White went on to build some extremely big tractors, which were very popular in their native country. They never came to Great Britain, and in 1991 AGCO, the parent company of Massey Ferguson, bought the brand and the name was discontinued.

Many of the Minneapolis Moline tractors that came to Great Britain during the war are now in the hands of some very lucky private collectors.

New Holland

New Holland was renowned for making harvesting equipment that was found on many farms in the 1970s and 1980s, and the deep red and yellow hay balers were a common sight. However, balers were just one of many machines made by the company: forage and combine harvesters were also part of the-line up, and in the late 1990s the name started to appear on tractor bonnets.

The company's history stretches right back to 1895 when, in New Holland, Pennsylvania, a mechanical genius called Abram (Abe) Zimmerman set up a blacksmithing shop and promptly started to produce agricultural equipment. The company was very successful, and Zimmerman soon started selling Otto stationary engines, which were made in Philadelphia, but soon decided to sell Columbus single-cylinder engines instead. Seeing these engines at close hand made him realise that he could produce a better engine. In 1900 he did just that; his engine was better in most areas and was also frost-proof thanks to his design of water jacket. Twenty-seven years later he was employing 225 people and the company was making various items of agricultural equipment.

The Sperry Rand Corporation bought New Holland in 1947 and started to concentrate on hay-making machines and developing a mower conditioner. Claeys was a major combine manufacturer in the 1960s, and New Holland invested in this company in quite a big way, making it an even bigger player in the harvesting area. Just under 40 years later, after many successful decades, Ford bought New Holland, having seen John Deere's success in not only producing tractors but also implements. This meant that customers could buy everything they needed from one dealer network. Unfortunately for Ford, the recession was biting and the company

BELOW The Ford 40 series was rebadged New Holland. *New Holland*

ABOVE The New Holland 8560 came onto the market in 1996.
New Holland

needed to get rid of something; it was decided that the tractor and implement side should be disposed of, and when Fiat offered to buy an 80% stake this suited both parties. Now the tractors on which Ford had built an excellent reputation were set to become New Holland.

The first New Holland tractors were rebadged Fords, and from a distance they looked identical as Ford's colour scheme of blue with black was retained. On the first few models 'Ford' could be seen written in small letters under the New Holland name, but the famous blue oval had been replaced with the one that Fiatagri had used for many years. Basildon, the plant that Ford had set up in the mid-1960s, continued to build these tractors, and in 1996 the 40 series, which was largely similar to the one Ford had introduced, celebrated the emergence of the 75,000th example of the model.

In 1999 Case joined the family, which was then known as Case New Holland (CNH). To prevent this new partnership becoming too big and powerful, both New Holland and Case had to discard assets from within their individual organisations. Case carried on as its own company, and is dealt with in its own section. In Great Britain the tractors being produced were called New Holland, while in other European countries they were identical in shape but painted in dark orange and carried Fiat badges. Fiat tractors had been sold in Great Britain in the past, but not in any great numbers, so using the New Holland name instead, which was more recognisable in the agricultural world, helped the tractors to be as successful as they had been under the Ford name.

New models rolled off the Basildon production line, which was just one of many factories where these blue and black tractors were built around the world, including Italy, Turkey, Brazil and America. The colour scheme remained the same, although the original blue became lighter and the black decals were changed to blue with a yellow stripe, giving the machines a much-needed brighter appearance. These subtle colour changes moved the tractors away from looking like just rebadged Ford models, and gave New Holland its own identity.

TM135 NEW HOLLAND
NEW HOLLAND
BB940

ABOVE New models came quickly as New Holland established its brand in the tractor market. This TS115 is fitted with a Kuhn mower. *New Holland*

OPPOSITE The new millennium saw the TW135 arrive from the Basildon factory. *New Holland*

In 2014 the Basildon plant celebrated 50 years of tractor production by launching a number of Golden Jubilee models; these were slightly different from the regular tractors with a more upmarket feel, including leather seats. Outside, the grille was finished in gold, as was the exhaust shroud, and the normal colour of blue was replaced by a much darker shade.

The current range of tractors are identified with the letter T. The smallest is the T4, which comprises four models, the N, F, Powerstar 4B and, the biggest, the Tier 4A. Some of these models feature the new deluxe VisionView cab and the latest common rail diesel engines, which are some of the cleanest available. The T5 range is slightly bigger with Tier 4A and B; the TD5 features a new concept in air-conditioning systems comprising nine vents positioned

TM190
New Holland

New Holland
TS135
HR 4003

conveniently around the cab for maximum driver comfort. The TD5 is also probably the most versatile model in the range, and is particularly handy for front loader work. The T6 models in the range include the Tier 4A and 4B, and both have the ECOBlue engines, which the company claims deliver a 10% saving on fuel bills. The cab is an extremely quiet environment for the driver at only 71dB, with Comfort Ride cab suspension, while the impressive expanse of glass gives a 360-degree view. The bigger T7 and T8 models complete the conventional tractor range. However, the biggest New Holland tractor is the four-equal-wheeled T9 series; this massive centre-pivot-steer range of tractors packs a power output of 360 to 564hp.

Henry Ford would probably be quite proud of New Holland. When he originally conceived the Fordson F it was as a tractor that would be available to farmers easily and would be simple to use. New Holland tractors today are popular, readily available and, looking at the new models on the market, very driver focused. Every effort has gone into their design to help the driver work in comfort and safety. Although it is a shame that the Ford name has disappeared from the agricultural industry, New Holland has a rich history that looks set to continue well into the future.

TOP LEFT **By the time the TM190 was on the market, New Holland had produced new decals, which moved the tractors on from the old Ford models.** *New Holland*

BOTTOM LEFT **The new colour scheme and decals allowed New Holland to step out from Ford's shadow. The TS135A seen here is using a Kuhn rotary harrow.** *New Holland*

ABOVE **To celebrate 50 years of production at Basildon, special tractors were introduced throughout 2014. These special edition models had a darker shade of blue and elements of gold on various components as seen on this T7 example.** *New Holland*

LEFT **The T4 is one of the smaller tractors in the T series.** *New Holland*

BELOW **A standard T7 model pulling a Horsch seed drill, which not only drills the seed but also applies fertiliser at the same time.** *New Holland*

TOP RIGHT **One of the bigger models, the T8, is seen here baling straw with a New Holland baler.** *New Holland*

BOTTOM RIGHT **The biggest tractor in the range is the four-equal-wheel centre-cab T9 model.** *New Holland*

NEW HOLLAND

NEW HOLLAND

Nuffield

As the Second World War gripped Great Britain there was a big push to feed the nation and tractors were in big demand; many came from America, but other than those there was really only the British-built Fordson. With the country almost on its knees, having a machine available to sell both in Britain and to export further afield appealed to the Government. William Morris already had British

LEFT **Field tests of the Nuffield tractor were carried out in Great Britain in 1946 at a time when tractors were in great need. This half-tracked version is fitted with a trailed plough.** *Nuffield and Leyland Tractor Club*

BELOW **This very early prototype was already looking like the finished product. The field test of these models were carried out in Pershore in 1946.** *Nuffield and Leyland Tractor Club*

TOP RIGHT **An unrestored Nuffield Universal fitted with a Perkins diesel engine...** *Ben Phillips*

BOTTOM RIGHT **...and a restored example.** *Ben Phillips*

NUFFIELD

NUFFIELD
TNX 281

ABOVE A Nuffield being shown to a salesman on a demonstration day.
Nuffield and Leyland Tractor Club

factories building cars, so when the call went out requesting a 'new' British tractor he responded, and by 1946 a prototype was ready. However, the country's metal shortage meant that it was two years before the new tractor, known as the Nuffield, saw the light of day. It was a simple machine fitted with a four-cylinder Morris engine and ran on petrol TVO (tractor vaporising oil). The country needed simple machines – this was not a time for over-complicated machines, and simple also meant cheap. The orange M4 tractor came in at less than £500, which was well within the reach of most farmers and, fitted with a three-point linkage and a PTO shaft, it was useful and up-to-date.

By 1950 a Perkins diesel joined the petrol TVO as an extra engine option; by now many farmers had seen the advantage of using diesel power, and Frank Perkins's company had perfected a good, reliable diesel engine, producing 38hp. This was a shrewd move, as five years later all but about 5% of Nuffield tractors sold were diesel-powered. The British Government was also happy, as 80% of Nuffields were exported, which had been the intention and certainly helped the economy. In the early 1950s a merger of a number of well-known British motor manufacturers took place, including Nuffield, to form the British Motor Corporation (BMC). In 1954 a new BMC diesel engine replaced the Perkins unit, a 3.4-litre engine producing a respectable 56hp.

In 1957 new smaller models, the Universal 3 and Universal 4, were introduced, with a 37hp BMC diesel engine; these were aimed at customers who needed a slightly smaller machine. The Universal 4 was by far the most popular; it featured independent power take-off and independent brakes, and was exported to nearly 80 countries.

The dawn of a new exciting decade saw new 1960s models developed with a new numbering system. The 3/42 had a BMC three-cylinder diesel engine, while the 4/60 was built in Cowley in Oxford and had a BMC four-cylinder diesel. Most Nuffield tractors so far were built in Birmingham, but a new factory was being built in Bathgate in Scotland. Geographically speaking it might seem odd to move a factory from the centre of the British Isles to somewhere quite far north, but the move was made to help the Scottish people recover from coal-mine closures in the area. It was also hoped that a number of specialist businesses would start up to supply the new tractor plant with parts. Unfortunately this did not materialise and Birmingham still made most of the components, which added transport costs.

The 1960s will forever be remembered for the Mini, either in skirt of car form, but few realise that a tractor can be added to the list. The

BMC Mini came onto the market in 1965; it was developed in Coventry by the Harry Ferguson research company, so it is not surprising that it looked similar to the Ferguson TE20. But although the latter was a great success, the BMC Mini was less so, probably due to the times, as the Mini was small when other tractors were getting bigger. The paltry 15hp BMC 950 diesel was quite feeble, but the tractor found its own niche in market gardens, smallholdings and, later, golf courses. What the Mini lacked in power it made up for in innovations: the gearbox was constant mesh, which made its nine forward and three reverse gears a joy to shift from one to another – I don't know of any other gearbox from that time to be so smooth.

If the Mini looked fresh, the bigger Nuffield tractors were getting to look old, and replacements were needed. The 3/45 and 4/65 came onto the

RIGHT 1964 saw the introduction of the Nuffield 10/42 diesel. *Ben Phillips*

BELOW The same year (1964) saw another introduction, the 10/60. *Peter Phillips*

ABOVE Nuffield tractors stood out as they were all bright orange in colour. *Peter Phillips*

RIGHT The BMC Mini was the smallest in the line-up. It was not capable of major farm jobs, but was perfect for smallholdings. *Ben Phillips*

market in 1967, boasting a number of new features to bring them right up to date. New hydraulics with increased pressure allowed them to lift heavier items, but the fuel tank was fitted in front of the radiator, which proved not to be a good move when tractors became prone to overheating. The whole tinwork was updated quite comprehensively and a fresh new appearance was given to the new tractors. The instrument panel was also new, as was the steering wheel; in all, the update really changed the ageing Nuffield tractors.

The new styling that the Nuffield 4/65 received brought the tractors up-to-date. *Ben Phillips*

The Mini also had an update and a number change, now being known as the 4/25. The lack of power that had dogged the original tractor was addressed with a new 1.5-litre diesel giving a 10hp increase to 25hp. A slightly more powerful 1.6-litre petrol engine was also offered, giving 28hp, but with most tractors the diesel version was the main seller. Very slight bonnet changes and a new silver decal running the length of it were the only visual differences made to the tractor.

Nuffield was no stranger to mergers, and in 1968 Nuffield BMC tractors became Leyland; the orange tractors were set to be replaced by a whole new range in two-tone blue, and a new chapter opened for this truly great British company.

TOP **The 4/65 being built on the Nuffield production line. The factory was at Bathgate in Scotland which had training facilities on site and good transports links. A neighbouring farm called Mosside provided an agricultural backdrop to many publicity photographs.** *Nuffield and Leyland Tractor Club*

ABOVE **Nuffields through the ages – the example on the left is a restored prototype, the model in the centre is a 1949 Nuffield M4 and on the right is a 1968 Nuffield 4/65.** *Ben Phillips*

Renault

Louis Renault was certainly a pioneer of the motor car in his home country of France but, not content with producing just cars, he set about manufacturing agricultural machines. By 1918 the Renault factory had experience of making tanks for the First World War, so it is not surprising that the first offerings looked rather tank-like, complete with tracks. Louis tested these tractors on his own farm in the French countryside, and was obviously keen to develop agricultural machines, seeing a real need for them in France. Soon wheels replaced the tracks and they began to look more like tractors as we know them today. With production going well in the Le Mans plant during the 1920s and 1930s, a new factory was planned, but the Second World War intervened; Louis Renault did not survive the war, so the Government took control, and continued to operate the Le Mans factory. By the 1980s it had grown in size, covering 21 hectares, with 7.5 of that total under cover. The factory was modern and well-equipped and featured the latest technology, following the similar philosophy of the company's car factories.

Renault tractors only really took off in Great Britain during the 1980s and, as with other foreign tractors, they tended to be found in certain areas of the country while rarely seen in others. The bright orange bonnet and wings contrasted with the dark grey of the rest of the tractor. The first examples seen in Britain were very square and angular, which was typical of French design in the 1980s.

By the late 1980s the most popular model was the Renault TX, and at the beginning of the 1990s the bigger TZ was introduced. These models ranged from 70 to 170hp and featured well-equipped cabs that Renault had christened Hydrostable in 1987. This cab suspension system was meant to bring more comfort with suspension seats, moveable steering wheels and full climate control; Renault really went all out to produce the most comfortable working environment for drivers. Also at this time electronics on tractors were becoming the norm, and Renault tractors had Tracto-control (TCE) on their hydraulics. Controlled by rotating dials and switches, this replaced the old-fashioned levers of the mechanical system and allowed much better and more precise control than the

BELOW **Renault tractors became established in Great Britain in the 1980s, and soon models like this 155.54 were a common sight on farms.** *Renault Agriculture*

ABOVE **The turning circle on Renault tractors was a big selling point for the company.** *Renault Agriculture*

RIGHT **The Renault Ares models were introduced in the late 1990s and were the company's last tractors.** *Renault Agriculture*

operator could achieve by pulling a lever. A further electronic package could be fitted called Tractorador, which helped to monitor work rate and control wheel slip.

Updated models in the late 1990s and early 2000s included the Ares, a model range that was probably the true successor to the TX and TZ. They looked similar, but featured much more rounded bodywork and had power outputs from 85 to 205hp. The cabs were also heavily updated: gone was the squareness of the dash, gauges and switches, and a far rounder appearance greeted the driver. There was also more space, which allowed a passenger to be present in the cab, even provided with a little seat. The visibility was improved with a large expanse of glass, and the slope on the bonnet helped with the view forward. Further improvements to the Hydrostable cab made Renault believe that it provided the most comfortable ride on the tractor market.

ABOVE **It was a shame Renault moved out of agriculture after so long a history as a tractor producer. Models like this TS16 are still around working on farms today.** *Jane Brooks*

In 2003 Renault sold a majority stake (51%) in the company to Claas, the German firm famous for producing combine and forage harvesters. Unfortunately this spelled the end of the Renault name in agriculture and, although production still took place at the Le Mans factory in France, from 2005 all tractors were badged Claas. The deep yellow paintwork was replaced with Claas's light green and white; although for a while the design was still Renault, the name had gone for good and Claas started to increase the worldwide presence of its product. Although Renault was quite popular in France and in certain other European countries, including Great Britain, it did not seem to have the same hunger for global growth as Claas.

Siromer

Compact tractors are very popular throughout Great Britain, and Siromer has certainly cashed in on this recent boom. In 1999 smallholder Jeff Howard was in need of a compact tractor, but found the choice in this area of agricultural machinery somewhat limited. He therefore embarked on a journey round the world looking for a solution, ending up at a factory in China, from where he began importing tractors. Instead of being shipped fully assembled, they were tightly 'flat-packed' in a metal frame, which kept costs to a minimum. When they arrived in Great Britain they were assembled and delivered to the customer; however, one customer, who was being shown how to use his Siromer, said he would like to build it himself, so since then the company has offered the tractors either assembled or flat-packed. More than 3,500 tractors have so far been sold, with quite a percentage being assembled by their new owners.

Pete Kelly, who was then editor of *Tractor and Farming Heritage* magazine, took a Siromer 404 on a 1,000-mile road trip around England, Wales, Scotland, Northern Ireland and the Irish Republic; this enjoyable journey took him 10 days and was featured in the magazine, proving the Siromer to be reliable and comfortable, even if he was exposed to the unpredictable March weather.

The Siromer models are split into three ranges. The Field range, the first models introduced, feature a power output between 20hp and 28hp. Having direct-injection diesel engines and flow-and-return hydraulics, these tractors are very capable, and turf or agricultural tyres can be specified.

The four-model EU range are road-legal tractors that came onto the market in 2009 with 16-35hp indirect diesel engines. To improve visibility a mirror is fitted, and added features such as a comfortable sprung seat and good hydraulic outlets via four quick-release connectors give this compact tractor the feel of a much bigger machine.

The biggest Siromer tractors come in the form of the CH range, which offers 25-50hp and can also be road registered. They have a more sophisticated shuttle gearbox and come in three models, the 254CH, 404CH and 504CH. As this range is a lot heavier at 2.5 tonnes, the front and rear axles are a lot stronger and a better wet braking system is offered. The CH range will also take all of the bespoke implements developed by Siromer, which are the same as those found on bigger tractors: mowers, balers and even a reversible plough in a size to match these mini tractors.

RIGHT Siromer has always kept its tractor designs up to date, adding to their popularity. *Clinkaberry Tractors*

BELOW Siromer tractors have been extremely popular in the 'mini' tractor segment of the market. This 304 is doing what it does best somewhere in Southern England. *Clinkaberry Tractors*

Siromer

Steyr

Considering the topography of Austria and the amount of countryside it is not hard to see why Steyr became a very popular tractor manufacturer.

On a September day in 1947 the first Steyr tractor was delivered to its purchaser. It was a 26hp Type 180 machine that was painted in green, and little did anyone think that it would still be going strong 70 years later. To complement the little green tractors, a year later the company began to produce implements, and hydraulic hitches were fitted on new models from 1950; they could also be fitted to all the earlier models. The mid-1950s saw a number of new models enter the market thanks to Austria becoming a free nation.

As with most tractor manufacturers, Steyr is known for two colours, red and white, a scheme that was adopted in 1966. Around the same time the company started to offer Perkins engines, which helped it build bigger tractors to enter a whole new market. During the 1980s the tractor market was going through somewhat difficult times, but Steyr managed to emerge largely unscathed. Then the company's future was secured when Tenneco, the parent company of American tractor giant International

ABOVE **The earliest Steyr tractor, the 180, was delivered to its first customer in 1947.** *Steyr*

BELOW **Steywr soon became known for its red and white colours; this 9094 was introduced in 1993.** *Steyr*

Harvester, bought it. Steyr's presence in Great Britain was further enhanced by the dealer Morris Corfield. Based in Shropshire, this firm had a number of lesser-known tractor brands on its books, and was soon selling a number of Steyr tractors in and around its sales area; farmers soon began to see how good this Austrian brand was.

The factory was kept busy building a number of Case models, and was engaged in developing other Case tractors when Case IH was bought by New Holland. Steyr was included in the sale and now became part of one of the biggest tractor brands in the world. Unfortunately in recent years the company's market share in Great Britain has declined, but back to the 1990s they were a popular tractor in certain parts of the country thanks to a certain Shropshire tractor dealer.

ABOVE 1983 saw the introduction of the Steyr 8090, and it was during the production of this tractor that the company began to sell in Great Britain. *Steyr*

BELOW This is one of the latest Steyr tractors: a CVT6165 model. The company is now owned by Case New Holland, and makes an important contribution to the Case side of the business. *Steyr*

TAFE

After a quick glance at any TAFE tractor, the observer might be forgiven for thinking it was an earlier machine made by Massey Ferguson. Tractors and Farming Equipment Ltd (TAFE) was formed in 1960 in partnership with Massey Ferguson, hence the reason why its products resemble MF's old models. In India TAFE has 25% of the tractor market and, with 150,000 tractors produced and sold every year, this firm is a major player in the agricultural market. It also still has close links with AGCO and together they export tractors to more than 85 countries throughout the world, Great Britain being one of them.

Tractors UK, based in Dorset, is the sole importer of TAFE tractors into Great Britain, and their everlasting appeal lies in one word – simplicity. Today most tractors rely heavily on electronics; even the smallest of tractors has an electric control unit to work everything, and when it goes wrong it normally involves a visit to a dealer to get it fixed. TAFE, however, has kept everything simple so the average farmer with a little knowledge probably has a chance to fix it. TAFE customers come from every area of agriculture – farming, equestrian and horticulture – and they all like the simplicity of TAFE tractors. Another reason why they are so popular is their familiar look; seeing a tractor that looks like an old reliable Massey Ferguson workhorse all adds to sales.

If the TAFE 35 Classic looks familiar, it has the same tinwork as the Massey Ferguson 35. This old design has stood the test of time well and still looks as good today as it ever did, although obviously there have been a few updates under the tin to comply with new laws. The engine is still a three-cylinder diesel and produces 37hp, but a Bosch inline fuel pump has replaced the CAV unit found on the original MF 35. The engine has been manufactured to meet strict emission laws, and more protection has been added around moving parts such as the fan and associated belts. The squarer lines of the TAFE 35DI imitate the tinwork of the Massey Ferguson 240/250 range, and it also has a 37hp three-cylinder diesel; both models had a roll bar fitted, as stipulated by law if the tractor does not have a safety cab. The TAFE 35DI OIB has a cab fitted, which was not carried over from the Massey Ferguson model on which it was based. The biggest models TAFE build are the 45DI and 45TDI, and these also feature familiar Massey tinwork. These bigger models feature a four-plate oil-immersed disc braking system, and all have

BELOW **You could be forgiven if at a glance you thought this was a Massey Ferguson 35, as the tinwork is identical to that model on this TAFE 35.** *Peter Phillips*

hydrostatic power steering and a live PTO. Other standard features on all TAFE tractors include eight forward and two reverse gears and a relatively comfy seat.

It is nice to see old Massey Ferguson designs reused on TAFE tractors, and this firm has given people the chance to own a new tractor that is simple to use and maintain and, more importantly, affordable. It has also kept alive the manufacturer of the old-style tinwork, and that in turn has helped the tractor preservation scene; these days if you need a bonnet, dash or fender for your old Massey Ferguson, the chances are that one is available.

ABOVE **Again based on a Massey Ferguson, the TAFE 450DI has brought four-wheel-drive tractors to people who otherwise could not afford them.** *Ben Phillips*

Turner

Wolverhampton in the Midlands was a great place for early industry; it was geographically situated in the centre of Great Britain and had ideal transport links via an extensive canal network. Because of this all manner of businesses thrived in the area, dubbed 'The Black Country'.

The Thomas Turner Company was founded in the mid-1800s in a tiny workshop behind a simple house in Wolverhampton. Outgrowing these humble premises, the firm moved to new ones, and it was here that they started to build cars. Car production was good until the First World War caused it to cease in favour of machine tools for the war effort, then after the conflict car production resumed. As well as cars, Turner Motor Manufacturing, as it was now known, was building engines for boats and heavy-duty winches. At the start of the Second World War Turner was again an important company, helping to provide vital aircraft parts, work that continued after the war. However, Turner was looking for something else to produce, and with demand for agricultural tractors at an all-time high it was too good an opportunity to miss.

The Turner Yeoman of England was introduced in 1949 and featured a diesel engine built by the company itself; such engines were rare in tractors at that time and Turner thought it was onto something good. This was a time when Great Britain needed good products to sell and export, and when the Turner tractor was conceived everything pointed to it being a big success, but unfortunately that was not to be.

The Turner certainly looked the part: two shades of green, with a yellow grille and wheels was a great colour scheme; it was also quite a big tractor, and looked well made. The V4 engine let out a fantastic exhaust note but problems with overheating manifested themselves almost immediately; these had been evident from when it was being tested. Other problems that supposedly arose regularly were head gasket failures and a fragile gearbox, in which third gear was prone to breaking. The steering box was originally designed for a car, and little thought had been given to the added strains on this part. If these problems were not enough to put off potential buyers, the price tag of nearly £800 was. At this time Fordson and Ferguson were the best-selling tractors in Great Britain and, with the price of these mass-produced machines cheaper and offering greater reliability, it is not hard to see why the Turner Yeoman of England failed. The company tried desperately to sell

them, and a Mark 3 was brought out with a redesigned cylinder head and gasket that would hopefully address the problems. Also by now a new larger radiator had been fitted to keep the overheating engine cool. In a final act to attract customers the price was reduced quite considerably, but even so it was still much more expensive than its rivals.

In 1955, after just six years, the Turner Yeoman of England went out of production with just over 2,100 having been sold. In hindsight this project was doomed for failure from the start; expecting a small firm from Wolverhampton to produce a tractor cheap enough to take on the might of Fordson and Ferguson was just too much. Today the company still exists, and makes transmissions for various agricultural machines; it is now owned by Caterpillar.

ABOVE **Extensive tests were carried out in the late 1940s which revealed a few problems. Most notable were cooling and head gasket issues.** *turnermanufacturing.org.uk*

OPPOSITE **A Turner undergoing tests, probably in Wolverhampton around 1949.** *turnermanufacturing.org.uk*

ABOVE The Turner was a big, heavy tractor capable of lifting a big three-furrow plough. *turnermanufacturing.org.uk*

RIGHT A cab could be fitted to the Turner, which at least kept the driver dry and semi-warm. *turnermanufacturing.org.uk*

LEFT Turner Yeoman of England adopted green and yellow colours for its tractor. *Ben Phillips*

ABOVE **The Turner tractor could also have spade lug wheels fitted instead of pneumatic tyres.** *turnermanufacturing.org.uk*

TOP RIGHT **Head gasket problems were evident in some engines, just one difficulty that beset this tractor.** *Ben Phillips*

BOTTOM RIGHT **With just over 2,100 Turner tractors having found homes, production stopped in 1955.** *Ben Phillips*

Valtra

ABOVE Valtra bought Volvo tractors, and the latter's influence can be seen here on the 8150. *AGCO Archives*

The Valtion Metallitehtaat, a Finnish metal works, shortened its name to Valmet just after the Second World War. Not until the early 1950s did the first tractor roll off the production line; called the Valmet 15, this small 15hp tractor was produced so that farmers could have something mechanical with which to replace their horses. Valmet tractors went on to be very successful, and many ended up in Brazil, where a factory was soon established. With Finland and Sweden being neighbours, it was not surprising that Valmet bought Volvo tractors and became the biggest Scandinavian tractor producer. In 1994 the Sisu Corporation took control of the company and was allowed to use the Valmet name for a set period of time. In fact, in the interim it was called ValtraValmet, Valtra being a name that had been associated with Valmet since 1963. Over the years a few Valmet tractors found their way into Great Britain, but not in great quantities. In 1998 the name Valmet was due to be replaced and it was widely thought that the machines would be called Sisu; however, this was decided against, as Sisu does not easily roll off the tongue, so the brand was known from then on as Valtra.

Although a large well-known company in certain territories, in 2004 the company was purchased by AGCO, the parent company of Massey Ferguson. This opened up a whole new future for Valtra tractors, as AGCO had the resources to promote the brand all over Europe.

Since then the tractors have had an ever-increasing presence in Great Britain with their distinctive bonnets, and have been recognised with a number of awards for innovative features. One was AutoComfort, an active suspension for the cab, and another was a system used on trucks whereby the exhaust cleans the gases emitted from the engine; since Valtra adopted this, nearly every other tractor manufacturer has followed suit. Valtra tractors are only made to order, and the customer can specify many of the specifications, even down to the colour. Other tractor manufacturers make them in one uniform colour by which they have been recognised for years – Massey Fergusons are red and grey, John Deeres are green and yellow, and so on – but a Valtra tractor can be in virtually any shade the buyer fancies. The Young Farmers in Finland asked for and got a pink Valtra, while another farmer wanted one to look like his cows; 106 Valtra tractors were also specially customised for the Finnish Defence Forces.

Tractors are not known for being fast, even though new ones can travel at quite a speed if needed. To prove that Valtras are not slow, in 2015 Juha Kankkunen, four-times world rally champion, travelled at 80mph in one and set a new world speed record for a tractor!

TOP During the transition between names, the tractors were known as ValtraValmet. This is a T850 and is using a four-furrow plough. *AGCO Archives*

ABOVE It was widely thought that Valtra tractors would be called Sisu when the original name Valmet was due to be replaced in 1998 but 'Valtra' was chosen instead. This T191 is using a Metsjo Vagnen grain trailer. *AGCO Archives*

RIGHT Valtra also makes smaller tractors that are more useful around more confined British farms. This N series Valtra is using a Hardi Boom sprayer.
AGCO Archives

TOP It will be noted that hardly any Valtra tractors have been pictured in the same colour – that's because the buyer can choose any colour. This T151 is green and is fitted with a Valtra 75 front loader. *AGCO Archives*

ABOVE Presumably this Valtra is an exact copy of the markings of its owner's cows that he is about to feed with this JF Stoll Feeder. *AGCO Archives*

TOP RIGHT In Britain, as in Finland, many Valtra tractors are found in the forestry side of agriculture. These Valtra tractors are lifting timber with a Cranab CRF4DT. *AGCO Archives*

BOTTOM RIGHT Some Valtra tractors such as this A series model are even small enough to fit inside farm buildings. *AGCO Archives*

VALTRA

VALTRA

Valtra has come a long way since the Valmet days under AGCO's ownership – and Valtra has also been good for AGCO. This S Series Valtra-Valmet is pulling a long Lemken plough. *AGCO Archives*

VALTRA
VALTRA VALMET
1200 KG
1200 KG

Zetor

This Czech company started in 1946 and took its name from the first three letters of an arms logo on its factory and the last two letters of the word tractor. The first model produced was the Z25, which had very advanced features including a diff-lock, independent brakes and a diesel engine. Most tractors of the era only had petrol engines fitted, so producing a tractor with a diesel put the company certainly a few years ahead of the competition. Also ahead of its time was the wider range of gears available compared to others on the market, all of which gave Zetor a great reputation.

In succeeding years more tractor models appeared, riding on the initial success of the Z25, and each new model featured further advances that would become mainstream after Zetor had introduced them – there were certainly a lot of clever people employed at the Czech factory! Not only were they technically sound, but during the 1950s and 1960s the designs were also good. The colour scheme of a light gold skid unit with orange/red tinwork and wheels bolted to it was an almost perfect combination. The models from those two decades were some of the very first to feature parts that could be fitted to every model in the range; known as 'unified', this meant that production costs were kept down and ordering spare parts was easier. This system would become normal for most manufacturers over the next few years.

Models such as the 3011 and 3045 were to prove popular all over Europe. The numbering system was not as random as it might appear; in the case of the 3011, the '30' indicated the basic tractor's power output, the first '1' meant that it was a two-wheel-drive model, and the final '1' that it was an universal tractor model. Even though Zetor tractors were often considered basic, this was largely not the case; during the 1960s they developed one of the first hitch systems controlled by the hydraulics.

In 1968 Zetor introduced the 8011 Crystal, which was probably its most iconic tractor, staying in production for almost 20 years. By now the tinwork was being squared off, including the cab, which

BELOW **The Zetor 3011.** *Ben Phillips*

RIGHT **The Zetor 3045 four-wheel-drive tractor was great value for money when it arrived in Great Britain.** *Ben Phillips*

PLEASE
KEEP OFF...
Zetor 3045

ZETOR CRYSTAL
ZETOR
12045

LEFT The Zetor 8011 Crystal was probably the most iconic of the company's tractors, and very sought after today by collectors. *Ben Phillips*

ABOVE During the late 1960s Zetors became more squared off compared to previous models, but this 3545 was still a great-looking tractor. *Ben Phillips*

was bigger than that of most other tractors on the market; in fact, the whole tractor was massive and built like a tank. Most farmers who owned one loved what it was capable of doing, which was pretty much everything, and even if the brakes were the weak link the tractor soon gained legendary status. This Czech company was always forward-thinking and saw the increasing need for tractors with decent cabs, so having fully-fitted cabs that were safe and comfortable was high on its list of priorities.

Throughout the 1980s and 1990s Zetor tractors remained strong, well built and cheap, the 5211 being typical of the time. During this era the threat to Zetor was from the eastern bloc, which was on the brink of disbanding, and the uncertain times did nothing for Zetor tractor sales in Great Britain. An unlikely saviour was John Deere, which let Zetor build some tractors for certain markets, including Africa, where John Deere could see potential.

When HTC bought a 98% share, the company was overhauled with many departments being restructured, the assembly line was modernised and many cost-saving areas were identified. Now Zetor had a new future, and a new tractor range was introduced, known as the Fonterra and Proxima. These new tractors ushered in a new era with fresh exciting features, most notably the electrohydraulic gearbox, which allowed a change of direction under load.

Today Zetor tractors are more popular in Great Britain than ever, helped by a good range of models that are fresh and up-to-date; they are certainly not a cheaply built tractor, a wrong assumption that dogged them for years. The biggest tractor in the range has seen the Crystal name reappear; this six-cylinder model has a simple-to-use cab and the engineering is as one would expect from this company – extremely robust.

Although Zetor tractors are a product of the Czech Republic, their links with Great Britain go back decades. Renowned for producing strong and reliable tractors, Zetors have been popular with many farmers here in Britain, and truly deserve their place in British agricultural circles.

ABOVE The Zetor Proxima is just one range of many that the company produces today.
Ben Phillips